The World of William Brown
Railways – Steam Engines - Coalmines

by
Les Turnbull B.A.(Hons), M.Ed., MNEIMME

Sponsored by the Alan Auld Group Ltd.

Above: William Brown's Grave in Heddon Churchyard Northumberland courtesy of Andy Curtiss

Published by

The North of England Institute of Mining and Mechanical Engineers
in conjunction with the
Newcastle upon Tyne Centre of the Stephenson Locomotive Society

ISBN 978-0-9931151-1-0

Copyright Les Turnbull Newcastle upon Tyne 2016

No part of this book may be reproduced in any manner without
the permission of the publishers except for quotations in articles and reviews

Acknowledgements

A work of this nature is not accomplished without the support and encouragement of many individuals and institutions. Foremost amongst these are my many friends at the North of England Institute of Mining and Mechanical Engineers: without the facility of the Mining Institute and the generous access to its unique collection of mining records, this book could not have been written. It is always invidious to mention names but it would be remiss of me not to record my particular gratitude to Jennifer Hillyard, Simon Brooks, Bill King, Alistair Brook and John Crompton. I am grateful for the substantial financial support provided by the Alan Auld Group and for the contributions of the individuals named on the back page. Colleagues at the Newcastle Centre of the Stephenson Locomotive Society have also been a constant source of encouragement. The staff at the Lit. and Phil., Newcastle City Library, Tyne and Wear Archives and Museums Service and the Northumberland Archive Office at Woodhorn have suffered my frequent questioning with professionalism and fortitude and for this I am very grateful. A debt is owed to Professor Mike Franklin of Swansea University who first introduced me to the letters of Elizabeth Montagu which have proved a very fruitful source. Gayle Richardson, library assistant in the manuscript department of the Huntington Library California, where the letters are housed, has responded to my many queries with great courtesy. I am grateful to the Duke of Northumberland for access to his important archive; and especially to the archivist at Alnwick Castle, Chris Hunwick, and his staff. An important part of the Brown archive is with the Society of Antiquaries of Newcastle upon Tyne and I am indebted to the society for their permission to use this material. Steve Grudgings has kindly given me access to his private collection and has been a constant source of encouragement and information. I am grateful to William Hays for access to the Gibson papers. Andy Curtiss of the Heddon History Society has for several years provided me with detailed local knowledge which has proved invaluable; and through this association I have had access to Anne Willoughby's research into the parish records. Richard Carlton and Alan Williams have been very supportive and kindly invited me to participate in the Carville excavations which was a unique experience. I trust the many others who have helped in so many ways but are not named will forgive me. Finally, Peter Jokelson and the staff of Newcastle Print Solutions have been a pleasure to work with.

Les Turnbull
Newcastle upon Tyne 2016

Contents

Foreword	p. 5
Chapter One: The Golden Triangle – William Brown's Homeland	p. 7
Chapter Two: The William Browns of Throckley Pitt House	p.21
Chapter Three: The Coal Trade from the River Tyne	p.30
Chapter Four: The Work of the Colliery Viewer – Brown's Profession	p.37
Chapter Five: Throckley Colliery	p.45
Chapter Six: William Brown's Waggonways	p.61
Chapter Seven: William Brown's Fire Engines	p.81
Appendix I: The Newcomen Engine in the Great Northern Coalfield	p.99
Appendix II: Family Connections – The Browns, Newtons and Watsons	p.111
Appendix III: The Archaeology of the Eighteenth Century Coal Industry	p.113
Bibliography	p.122
Index	p.122

Illustrations

Fig. 1	Extract from William Gibson's Plan of the Northern Coalfield	p. 7
Fig. 2	Newcastle Central Station in 1858	p. 8
Fig. 3	Part of a Survey of the River Tyne in 1670	p.11
Fig. 4	West Brunton Waggonway to Scotswood	p.13
Fig. 5	Holiwell Reins Colliery circa 1767	p.15
Fig. 6	Part of Isaac Thompson's Plan of Denton 1754	p.16
Fig. 7	Major Waggonway Routes to Lemington in 1767	p.17
Fig. 8	An Eighteenth Century Coal Pit	p.19
Fig. 9	Hedley's Wylam Way	p.20
Fig.10	A Newcomen Engines of about 1730	p.27
Fig.11	Wilson's Plan of the River Tyne 1754	p.30
Fig.12	A Newcomen Engine for Pumping and Water Wheel for Haulage	p.39
Fig.13	Bell's Map of Throckley Royalty in 1781	p.44
Fig.14	Mining on Throckley Fell in circa 1785	p.47
Fig.15	Holiwell Reins Staith at Lemington	p.48
Fig.16	Walbottle in 1767	p.53
Fig.17	Walbottle Dene in 1769	p.55
Fig.18	A Miner and his Lass	p.58
Fig.19	Cartouch from John Gibson's Plan of the Collieries in 1787	p.61
Fig.20	Part of Gibson's Plan of 1787 showing Waggonway on Wearside	p.61
Fig.21	Throckley Waggonway in 1755	p.62
Fig.22	The Bridge Over Walbottle Dene	p.63
Fig.23	Throckley Waggonway in 1769	p.64
Fig.24	The Main Way at Carville	p.67
Fig.25	The Throckley and Walbottle Waggonway in 1771	p.68

Fig.26	The Two Bridges at Walbottle	p.69
Fig.27	False Rails at Lambton 'D' Pit	p.70
Fig.28	Re-used Ships Timbers	p.70
Fig.29	Walbottle Colliery Village circa 1767	p.72
Fig.30	The Wash Hole at Newburn	p.72
Fig.31	Wash hole on the Willington Waggonway	p.73
Fig.32	The Chaldron Waggon	p.74
Fig.33	Black Close Colliery circa 1750	p.75
Fig.34	Newburn Village circa 1767	p.76
Fig.35	The Waggonmen	p.76
Fig.36	The Throckley and Walbottle Staithes at Lemington	p.77
Fig.37	West Longbenton Fire Engines in 1762	p.80
Fig.38	A Newcomen Engine at Tanfield Lea in 1715	p.82
Fig.39	Newcomen Engines in the Lower Ouseburn Valley	p.83
Fig.40	Part of Burleigh and Thompson's Plan of the River Wear	p.84
Fig.41	Old and New Technology at Heaton Banks Colliery	p.85
Fig.42	Engines at West Longbenton in 1749	p.86
Fig.43	Section of the Strata in the Tyne Basin by William Oliver 1761	p.87
Fig.44	Hartley and Seaton Delaval in the late 1750s	p.89
Fig.45	Shiremoor Colliery in 1790	p.90
Fig.46	Byker Old Colliery	p.91
Fig.47	Byker Colliery circa 1755	p.91
Fig.48	Byker St. Anthony's Colliery circa 1790	p.92
Fig.49	Walker Colliery Engines in 1780	p.93
Fig.50	Walker Colliery in 1763	p.94
Fig.51	Brown's Development of Walker Estate circa 1780	p.95
Fig.52	Gibson's Plan showing Willington in 1787	p.97
Fig.53	Wallsend 'A' Pit – Probably William Brown's Last Job	p.98
Fig.54	Brown's List of Newcomen Engines	p.99
Fig.55	Elswick Colliery circa 1755	p.106
Fig.56	Bushblades Engine of 1730	p.106
Fig.57	Bushblades Colliery in 1770	p.107
Fig.58	Friar's Goose Colliery in 1753	p.108
Fig.59	Blackett's Jarrow Colliery	p.110
Fig.60	Washington Estate in 1750	p.110
Fig.61	The Newcastle Miner	p.112
Fig.62	Extract from the Journal of Sir John Clerk April 1724	p.112
Fig.63	A wheeled sledge or tram	p.113
Fig.64	Underground at Walbottle Colliery	p.113
Fig.65	Bigges Main Village in 1858	p.114
Fig.66	East Benton Sinkers Medal presented to Matt. Tubman in 1786	p.114
Fig.67	The Steel Mill	p.115
Fig.68	The Ventilation Apparatus at Byker Colliery in 1724	p.115
Fig.69	Buddle's diagram of Coursing the Air	p.116
Fig.70	Cog and Rung Gin	p.116
Fig.71	The Whim Gin	p.117
Fig.72	Wallsend 'C' Pit	p.117
Fig.73	Wallsend c.1782 showing the beginnings of Wallsend Colliery	p.118
Fig.74	Remains of Wallsend 'B' Pit – possibly Brown's work	p.118
Fig.75	The Waggonway	p.119
Fig.76	The Coal Shoot	p.119
Fig.77	Lidar Survey revealing an Eighteenth Century Mining Landscape	p.120
Fig.78	The Coal Spout	p.124

Foreword

On September 25th 1825, the normally quiet market town of Darlington was in turmoil. Large crowds cheered George Stephenson on the footplate of 'Locomotion No.1' as the steam engine hauled a train of coal waggons packed with eager passengers en route to Stockton. The occasion was the opening of the world's first public railway. Five years later Stephenson would enjoy the same enthusiastic acclaim while at the controls of 'Northumbrian', another of his locomotives. Then the occasion was the opening of the Liverpool and Manchester Railway, the world's first inter-city line. To many historians these events marked the beginning of the railway age and a revolution in transport which was to change the world. However, the diaries of travellers in the early eighteenth century reveal that our fascination with railways and steam engines has a much longer pedigree. When, in April 1724, Sir John Clerk journeyed south from Midlothian to see Richard Ridley's waggonway to the port of Blyth and his three Newcomen engines at Byker Colliery on Tyneside, he was only one of several dignitaries lured by the appeal of these new technologies. In the following year, the Earl of Oxford was excited by the massive steam engine at Flatts Colliery to the west of Chester-le-Street and his chaplain wrote enthusiastically about the two railways from the River Wear which they later encountered on their journey northwards to Newcastle upon Tyne. Their enthusiasm was shared by the celebrated antiquarian Dr. William Stukeley who, in September 1725, was amazed at the civil engineering works newly erected on Sir Henry Liddell's waggonway from Tanfield to the River Tyne and by Lord Lowther's 'famous fire-engine' at Whitehaven. Nor was this enthusiasm for steam engines the sole preserve of men. In June 1766, one of the leading lights of London's literary scene, the queen of the bluestockings Elizabeth Montagu, was captivated by the massive steam engine built by William Brown to drain her colliery at East Denton, which is now part of the western suburbs of Newcastle.

Because of the development of the Newcomen steam engine for pumping water from mines, and the construction of wooden railways to transport coal to the riverside for shipment primarily to London, the coal industry in Northumberland and Durham was able to expand fourfold during the eighteenth century, an event of immense economic significance for the whole country. By providing the essential supply of domestic fuel to the capital, the Great Northern Coalfield enabled London to grow into what many regarded as the greatest city in the world; and the wealth generated by the coal trade transformed the North East, and particularly the regional capital Newcastle upon Tyne, into one of the most prosperous regions in Britain. The Northumberland and Durham Coalfield became popularly known as the Great Northern Coalfield because of its importance as the largest and most technically advanced mining region in the world; and its railways, referred to by foreigners as 'Newcastle Roads', were one of the wonders of the age. William Brown of Throckley Pitt House, six miles west of Newcastle, was at the centre of these developments.

Brown's papers, together with those of his assistants William Gibson, Christopher Bedlington and George Johnson, are an important source of detailed information about the world's largest coalfield in the mid eighteenth century. The correspondence of his neighbours Edward and Elizabeth Montagu give a wider perspective from a family at the centre of the nation's cultural events. Like many of the region's coalowners, the wealth generated by the Montagu's commercial interests on Tyneside enabled them to own property in the capital and play a prominent role in London society. These primary sources, together with the archive of the Duke of Northumberland, provide the principal evidence for this study. These documents shed light upon the vibrant industrial scene on Tyneside, a region dominated by the coal trade. They reveal the illegal machinations of the owners to rig the market at the expense of the people of London and their continuous financial worries as their mines swallowed up large sums of money before disgorging any profit. They show that the coalmine was a lottery – an enterprise constantly under threat from water, fire, adverse geology and industrial action. They highlight the independence and tenacity of the miners eager to maximise upon their scarcity value as skilled workers in an expanding industry. Like the workers in today's oil industry, these were men who provided the nation with its essential fuel supplies; they worked hard amid constant danger, earned substantial sums in comparison with other workers, and knew how to enjoy life to the full with good food, plenty of liquor and the occasional day off at the races. The documents also illustrate the complex and diverse role of the colliery viewer, that multi-skilled engineer who advised upon the direction of the mine, working at one level with the mining partnerships and royalty owners and at the other with the men at the coal face.

Of particular interest is the detailed documentary evidence which survives for one of William Brown's railways, the Throckley and Walbottle Way; and this evidence has been enriched by the recent excavations of the Willington Way at Carville, where remains of a wooden waggonway of international significance were discovered by the Newcastle Archaeological Practice in 2013. This data enables a very comprehensive account of a late eighteenth century railway to be written for the first time. It illustrates that these important arteries of the coal trade were complex railways which are not to be demeaned as simple tram roads. The waggonway was expensive to build: its cost often amounted to a quarter of the capital outlay for the colliery. The civil engineering works on the railway – the bridges, embankments, cuttings and staiths – were a source of pride to the locals and wonder to visitors from elsewhere in Britain and overseas.

But William Brown was primarily remembered by his contemporaries as a builder of steam engines: not the travelling engines of Stephenson's era but stationary engines for pumping water from the mines. His activities were not confined to the North East of England for coal owners in other parts of England, as well as in Scotland and Ireland, were keen to draw upon his mastery of the Newcomen engine. He was associated with many of the great engineers of the eighteenth century, men such as John Smeaton, James Brindley and Abraham Darby; his clients included dukes and earls as well as the lesser gentry. Because of his skill as an engineer and entrepreneur, William Brown rose from being a partner in a small landsale colliery to becoming one of the major coalowners on Tyneside; and his grandson would become High Sheriff of Northumberland in 1827. The family even acquired a coat of arms.

In February 1782, while William Brown was being laid to rest in Heddon churchyard, the infant George Stephenson, barely six months old, was cradled in the family's humble home at High House Wylam, a short distance to the west. From such modest beginnings, George would rightly rise to fame as one of the most accomplished engineers of the early nineteenth century, while the man who had been widely regarded as the 'father of the coal trade' because of his engineering skills would sink into obscurity. Such has been the pernicious verdict of history upon William Brown which hopefully this study will do something to rectify.

Chapter One: The Golden Triangle – William Brown's Homeland

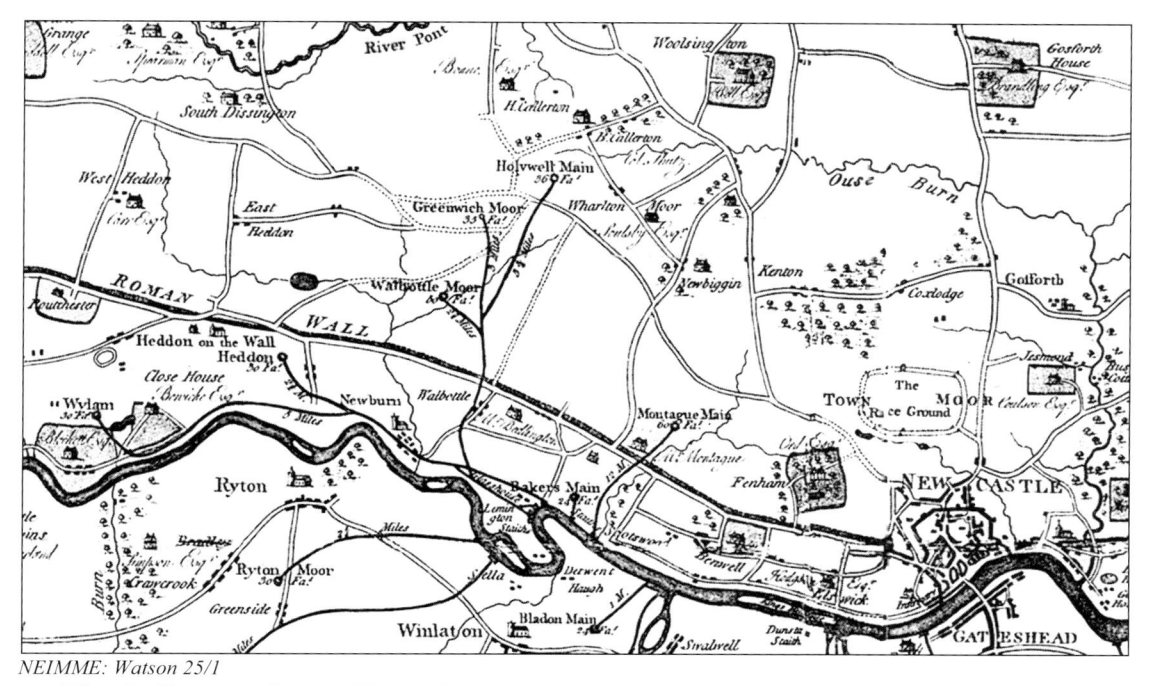

NEIMME: Watson 25/1

Fig. 1: Extract from Gibson's Map of the Great Northern Coalfield in 1787

The triangle of land to the west of Newcastle, with the River Tyne between Elswick and Wylam as its base and the airport at Woolsington as its apex, is a place of pilgrimage for railway enthusiasts. This area is remarkable for the number of distinguished engineers who worked in the coalmining industry during the early nineteenth century and made a major contribution to the development of both the steam locomotive and the iron railway. The best known is George Stephenson, who was born at High Street House to the east of Wylam village, on 9th June 1781, and spent his early working life in the small mining communities of Callerton, Dewley and Walbottle to the north of Newburn. But there were others who are also deservedly remembered as pioneers of the steam locomotive and iron railway.

William Hedley moved to Wylam from Newburn on becoming the viewer of Christopher Blackett's colliery. In 1808, Hedley rebuilt the wooden waggonway from Wylam to Lemington as an iron plateway; but his principal achievement was to solve one of the major problems exercising the minds of railway engineers at that time. In 1813, he demonstrated that the adhesive weight of the locomotive alone was sufficient to move a load of coals along an iron railway without resorting to toothed rails, chains or mechanical legs to prevent the wheels slipping.[1] In the following year, Hedley built two of the best known early locomotives Puffing Billy and Wylam Dilly for the Wylam line: Puffing Billy is now preserved in the Science Museum in London and Wylam Dilly in the National Museum of Scotland in Edinburgh. Jonathan Forster, a member of a distinguished family of waggonway wrights based in Newburn Hall, and Timothy Hackworth, the foreman smith at Wylam Colliery, were Hedley's assistants. In 1825, Hackworth became chief engineer to the Stockton and Darlington Railway and his engine works at Shildon now forms part of the National Railway Museum.

---

[1] John Blenkinsop used toothed rails on the Coxlodge Way, William Chapman used chains on the Heaton Main Way and William Brunton used an engine with mechanical legs on the Newbottle Way.

Another distinguished engineer, Robert Hawthorn, was born in Newburn and worked for fifty years (1789-1838) at the Duke of Northumberland's Walbottle Colliery, where he trained both George Stephenson and William Locke. Subsequently, their sons, Robert Stephenson and Joseph Locke, would make major contributions to the development of the railway network in Britain and the wider world beyond – indeed Robert Stephenson, the first engineer to be buried in Westminster Abbey, is regarded by many as the greatest engineer of Victorian times. In 1817, Robert Hawthorn established an engine works in Forth Street, Newcastle, for his sons Robert and William. Later, in 1824, the South Street works of George and Robert Stephenson was opened nearby. After more than a century of building locomotives, the two companies amalgamated in 1937 and the new firm continued to build locomotives till the end of steam on Britain's railways. [2]

Author's collection

Fig. 2: Newcastle Central Station in 1858

The skills of these early nineteenth century railway engineers were nurtured in the coal industry of the golden triangle; but their work is more than a matter of local or even national interest. The development of the iron railway, powered by steam locomotives, was one of the greatest technical achievements of the Victorian age, since it enabled heavy goods and people to be moved over long distances; and the fact that engineers from Britain exported this technology to Europe, Asia, Africa and the Americas gives their achievement an international significance. Although steam locomotives have been replaced by diesel and electric traction, and the iron railway has changed considerably in the past two centuries, the standard gauge established by the Stephensons is still used today by over sixty percent of the world's railways including the new high speed lines of Japan, Europe and China. The photograph above (fig.2) shows a locomotive standing at the west end of Newcastle Central Station in 1858. It was built for the Khedive of Egypt at Stephenson's factory which is a reminder of the international dimension of the work of the railway engineers from the North East of England. This engine is now a treasured exhibit in the Egyptian Transport Museum at Cairo. Hedley, Hawthorn, Hackworth and Stephenson are all recognised as pioneers and their story is rightly an

---

[2] Nor should it be forgotten that the mighty industrial empire of Lord William Armstrong, which stretched the length of the river from Elswick to Scotswood and was renowned for its guns, tanks and warships, diversified into building railway locomotives after the end of the First World War.

established part of railway history; monuments have been erected to their memory and biographies written about their achievements. But earlier, an equally talented group of engineers had worked with distinction in the same area. Unfortunately, their pioneering work has been largely forgotten: neither biographies nor monuments record the achievements of Richard Peck of Newbiggin, Christopher Bedlington of Walbottle and William Brown of Throckley.

Richard Peck was well established as a colliery viewer when, on 3rd November 1735, his eldest son Joseph was enrolled into the powerful Hostman's Company, which controlled the coal trade from the River Tyne. Richard leased collieries on Whorlton Moor, in Newbiggin and in Jesmond; but he was better known as a consulting engineer to many of the leading colliery owners of the early eighteenth century. In 1730, for example, he advised Lady Clavering on the development of Bushblades Colliery recommending the installation of a Newcomen engine to drain the mine and the building of a waggonway to Pickering Nook, the junction with the Derwent Valley Way, to get the coal to the London market. Peck was acknowledged as an authority on waggonways and Newcomen steam engines. In building the waggonway for Walker Hill Colliery, he inadvertently established the gauge which was to become the standard for the world's railways for this line was later used by Stephenson's employers, the Grand Allies, for their collieries at Heaton Banks and West Longbenton. His view book also reveals his involvement with railways for Jesmond, Hartley, Winlaton and Felling collieries; and for pumping engines at several collieries including Jesmond, Heaton and Byker on the western rim of the Tyne Basin. He advised the Earl of Carlisle on the development of Coale Fell Colliery in Cumberland. Nor were his interests confined to coalmining for he was associated with the Newcastle hostmen John Rogers and John Ord in the development of leadmines in Weardale. He died in October 1746 but his business interests were continued by his three sons. [3]

Christopher Bedlington and his brother William followed in the footsteps of their father, the viewer William Bedlington Senior of Shield Row, County Durham. Both sons were apprenticed to William Brown and two of Christopher's view books survive in the Mining Institute at Newcastle showing that he worked extensively throughout the northern coalfield and beyond. There are numerous references to steam engines and in 1766 he provided a detailed account of the building of the large 'fire engine' at Benwell to re-win the drowned colliery. Christopher has a claim to be the builder of one of the world's first passenger railways, the underground line from Scotswood to East Kenton Colliery, known locally as 'Kitty's Drift'. Work began on this railway beneath Elizabeth Montagu's land in 1796. Besides carrying coal, it was used by tourists wishing to explore a coal mine without experiencing the danger and inconvenience of being lowered down the shaft in a wickerwork basket. Akenhead's guide to Newcastle, written by the distinguished antiquarian the Rev. John Hodgson in 1807, suggested that this railway journey was popular with visitors at the beginning of the nineteenth century.[4]

But, the most important of these three eighteenth century engineers, was William Brown of Throckley Pitt House. He was recognised by his contemporaries as the

---

[3] North of England Institute of Mining and Mechancal Engineers (NEIMME): Peck/1; Hodgson, J.C., 'Richard Peck, an Eighteenth Century Coal Viewer', Archaeologia Aeliana, Third Series, Volume 8, p. 151-154; Raistrict A., The Steam Engine on Tyneside', Trans. Newcomen Society Vol.17 (1936 – 7).

[4] NEIMME: East 10A and 10B. Akenhead, Picture of Newcastle upon Tyne (Newcastle 1807), p.180.

leading authority on Newcomen 'fire engines' and he was also an acknowledged expert on early railways. His advice was sought by fellow engineers of the calibre of Carlisle Spedding, Lord Lowther's agent at Whitehaven in Cumbria, and Nicholas Walton and William Newton, the principal viewers to the Grand Allies, the largest consortium of coalowners in the world. The Bedlingtons, Gibsons, Barnes, Johnsons, Watsons and Chapmans, all distinguished families of mining engineers in the Great Northern Coalfield, were part of William Brown's professional network.[5] His stature amongst his contemporaries was such that he was a guest at the home of the great ironmaster Abraham Darby when he visited the famous works at Coalbrookdale. Furthermore, the list of coalowners hiring his services reads like a page from Burke's Peerage. Among his clients were the dukes of Northumberland, Hamilton and Portland; the earls of Carlisle, Holderness and Darlington; the Lord Crewe and Greenwich Hospital estates; the Ridleys, Blacketts, Delavals, Montagu's, Milbankes and Lambtons. Nor was his work confined to the Great Northern Coalfield: he had contracts in Nottinghamshire, South Yorkshire, Scotland and Ireland. Sadly, the name of William Brown does not appear in history books and the only monuments to his achievements are the grassy footpaths marking the routes of his forgotten waggonways.

The early history of the golden triangle, William Brown's homeland, begins between 360 and 250 millions years ago when violent climatic changes caused the deposition of several horizontal coal seams in what we now know as the North East of England. Subsequent massive earth movements fractured and tilted these coal seams to produce the complicated geology of the earth's crust which became the workplace of the Northumberland miner. The golden triangle is an area in which the tilting of the earth's surface has caused the principal coal seams to outcrop; and because the coal was easily accessible mining has taken place extensively since at least medieval times. Much of this early activity was in fact quarrying and it is interesting to note that the estate accounts of the Earls of Northumberland, lords of the manor of Newburn, record that the mines at Butterlaw were unoccupied due to the exhaustion of the surface deposits. But shallow shaft mining in the vicinity of the outcrops was also extensive; and a survey of Newburn Manor in 1613 records that 'there are no woods or underwoods of any value now left within the manor for that they have been greatly wasted and destroyed by James Cole and others for making steythes and timbering for cole pits'.[6]

The trade in coal from Northumberland and Durham began in the Middle Ages but the first great expansion of this important industrial enterprise took place during the reign of Elizabeth I. By the middle of the sixteenth century, supplies of wood, the traditional fuel used in both the home and industry, were becoming depleted and this stimulated a demand for coal. The subsequent growth of the population, particularly in London, together with the expansion of industry in the south eastern counties, acerbated the problem by adding to this demand. Coal is widely distributed throughout England except in the South East where the bulk of the population lived; but, because of its weight and the inadequacy of land transport in Tudor times, it was a difficult

---

[5] William and John Chapman sought Brown's assistance with the development of their collieries at Byker St. Anthony's and Wallsend; William and Christopher Bedlington and the Gibson brothers John and William were among his apprentices; Brown himself may well have served under Amos Barnes whose son Jonathan Barnes Junior subsequently became a member of Brown's consultancy.

[6] For a comprehensive discussion of early mining in the area see Jennifer Morrison 'Newburn Manor: an analysis of a changing medieval, post medieval and early modern landscape in Newcastle upon Tyne', Chapter Four, Durham University Masters thesis 2007, which is available as an e-thesis.

commodity to transport. Indeed, the cost of transport generally exceeded the cost of production if coal had to be moved more than about three miles overland. Furthermore, the market required large coal and the friable nature of the mineral meant that it had to be handled with care to retain the top market prices. Road transport was unsuitable. However, the coal seams in Northumberland and Durham were near to the rivers Tyne and Wear, which provided access to the sea and a route by water to London and the South East of England. The merchants of Newcastle took advantage of this opportunity and the coal trade increased from about 35,000 tons in 1550 to about 400,000 tons by 1625. The collieries in the area of the golden triangle made a major contribution to this trade. Figure four shows the concentration of staiths in the area between Lemington and Elswick on the western fringes of Newcastle in 1670.[7]

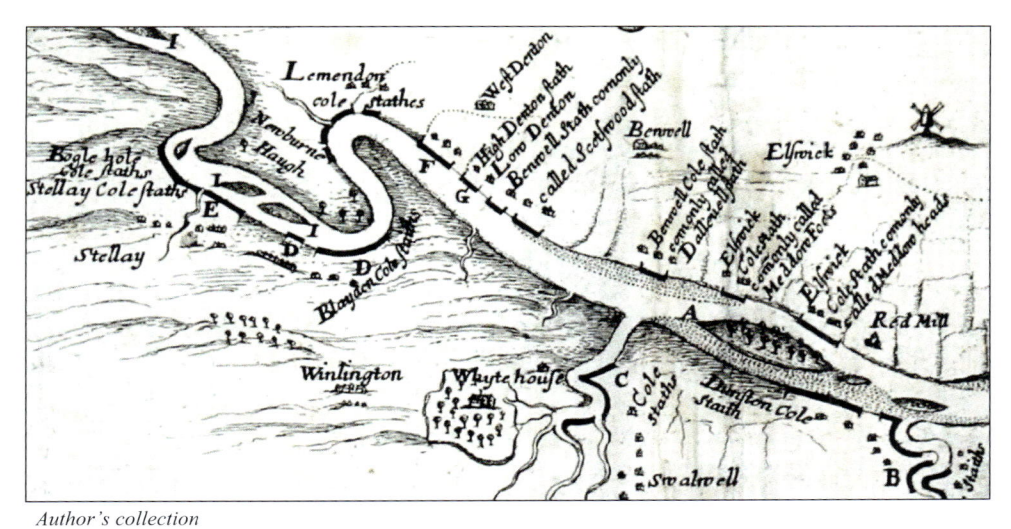

Author's collection

Fig. 3: Part of a Survey of the River Tyne in 1670

The trade from Tyneside continued to expand and by the time of Celia Fiennes' visit in July 1698 over a million tons was being shipped each year from the river. In July 1698, this intrepid adventurer travelled the road from Carlisle on horse back. The summer's day was scorching hot as she rode through Hexham to Corbridge and along the north bank of the River Tyne towards Newcastle. When she approached the town she saw an 'abundance of Little Carriages wt a yoke of oxen and a pair of horses together, wh is to Convey the Coales from ye pitts to ye Barges on the river'. She noted that 'this Country all about is full of Coale, ye sulphur of it taints ye aire and it smells strongly to strangers. Upon a high hill 2 mile from NewCastle I could see all about the Country wh was full of Coale pitts'.[8] Her viewing point was near the Roman fort at Benwell and she was describing the centre of the Tyneside coalmining industry at the end of the seventeenth century. The little carts were known as wains which were drawn by two horses and a pair of oxen. They carried about 17.5 cwt of coal often along specially built roads. The route of the wain road from Whorlton Colliery can be traced today from Westerhope golf course, along Hillhead Road and down Union Hall Road to Lemington staiths. A short distance to the east another wain road ran from the coal pits on the Earl of Carlisle's Newbiggin estate, along Newbiggin Lane and Beaumont Terrace, to Denton

---

[7] TWAMS: DX1381/1. Produced to accompany a report of the Conservators of the River Tyne 1670. See Brand History of Newcastle Vol .II. A – Kings Meadow, B – Team staiths, C – Derwent staiths, D – Blaydon staiths, E – Coleburn staiths, F – High Denton staiths commonly called Toppinore, G – High Denton staiths, I – Islands. Off the map was Bohemia staith 500 yards west of Newburn Church, Redheugh and Rock staiths in Gateshead.

[8] The text of Celia Fiennes journey from Carlisle to Newcastle and is available at Visions of Britain: Celia Fiennes.

Square and on to the River Tyne at Bell's Close. It is likely that Newburn Lane, leading from Throckley Moor downhill to Lemington, originated as another wain road. Curiously, Celia Fiennes does not mention waggonways, although the great Tyneside industrialist Sir William Blackett operated one from East Kenton Colliery to the River Tyne at the time of her visit.[9]

Waggonways, along which one horse was able to pull a waggon containing up to 53 cwt, were introduced at the beginning of the seventeenth century as a more efficient method of carrying coals; and by the end of that century waggons had replaced wains at the major collieries. These were the only collieries which had a sufficient volume of seasale trade to justify the large capital outlay necessary to build a railway. In 1714, Peck provided an estimate for the development of a colliery on Throckley Moor, about three miles from Lemington. The document is interesting especially because it illustrates the cost of the waggonway in proportion to the other development costs of the colliery. The proposal was to drive a drainage drift 700 yards long, presumable westwards from Walbottle Dene, and to sink three pits each 14 fathoms deep. The equipment needed is costed: roules, standers, hacks, wedges, shovels and candles. Twenty houses were to be built for the men at a cost of £200 in all. The wain trunk, the shoot for loading coals onto the keels at Lemington staith, cost £175. The sea-going vessels, known as colliers, were unable to travel beyond Newcastle because of the low medieval bridge and keels were used to ferry the coal downriver. The first option for transport was a wainway, priced at £50, which was to be provided with gates where it crossed the fields to prevent animals straying. The alternative was a waggonway but the capital outlay was very large: at £1,034 the expense was over twenty times that of the wainway and represented almost a doubling of the capital needed for the development of the colliery.[10] There is no evidence that this waggonway from Throckley Moor was built.

Richard Peck's papers also record an agreement between the landowner Sir Robert Hesilrige and the engineer John Wilkinson, dated August 1715, for the development of a colliery in Brunton, which was to have a waggonway linked to Sir William Blackett's line from Kenton Colliery. Blackett's line ran down the east side of Denton Burn to Scotswood.[11] Whether Wilkinson's branch line was built is not known. Kenton Colliery was drowned by 1717 but afterwards the railway may have been used for Blackett's colliery in Fenham. Interestingly, a map in the Watson papers (fig.4), which dates to later in the century, charts the route of an 'old Waggon Way' running due south from pits near the boundary of West Brunton and Dinnington (the site of Newcastle Airport) to Kenton Bank Foot, where an embankment is visible near the Twin Farms public house. The line then climbed up the east side of Newbiggin Dene to the plateau of Blakelaw. Throughout the eighteenth century, there was considerable interest in

---

[9] In 1688 Newcastle Council granted Sir William wayleave to carry coal from the south east part of Kenton, across the Town Moor, to staiths at Scotswood. Sir William was head of the more important branch of the Blackett family (not the Wylam branch) who resided in palatial accommodation within Newcastle town walls. Besides his coalmining interests, he was very active in the development of the lead mining industry especially in the Allendales. This branch of the family is remembered today in the name of Blackett Street in central Newcastle.

[10] NEIMME: Forster 1/5/21. Richard Peck was mining at Whorlton Moor in 1724 and at both Whorlton Moor and Newbiggin in 1738. His son Joseph was enrolled into the Hostman's Company in 1735. Richard died in 1743 (Archaeologia Aeliana 3rd Series, Volume 8, p. 151). Another son, William, took over his mining interests in Newbiggin and Holiwell Reins in partnership with Robert Beaumont.

[11] NEIMME: Forster 1/5/53. The agreement stipulates that 'Each waggon to be the same Gauge That Kenton Waggon is now of' but unfortunately it does not give figures. John Wilkinson, an important Newcastle hostman, was the partner of Sir William Blackett in the development of collieries at Felling and Carr Hill both of which were served by waggonways.

Brunton estate because the valuable High Main coal was only about twenty fathoms below the surface but water was a major problem. William Brown's survey of Brunton in 1752 recorded that 'all of these collieries are very disagreeably situated there being no level to get to drain them unless brought at a great expense from the River Tyne or the water drawn by fire engines; and this last method has the experience of being very expensive'. However, a later plan by William Brown, dated April 1763, shows four coal pits mining the High Main Seam near the boundary with Dinnington. These were the pits served by the West Brunton Way which was probably Brown's work.[12] After leaving Brunton the line followed the western boundary of Kenton estate to the Newcastle – Carlisle turnpike road and then ran down the west side of Denton Burn to staiths in East Denton probably using the existing waggonway for that colliery. Kenton Colliery re-opened in the late eighteenth century and the Kenton Way joined the Brunton Way a short distance before the turnpike road.[13]

NEIMME: Watson 20/4

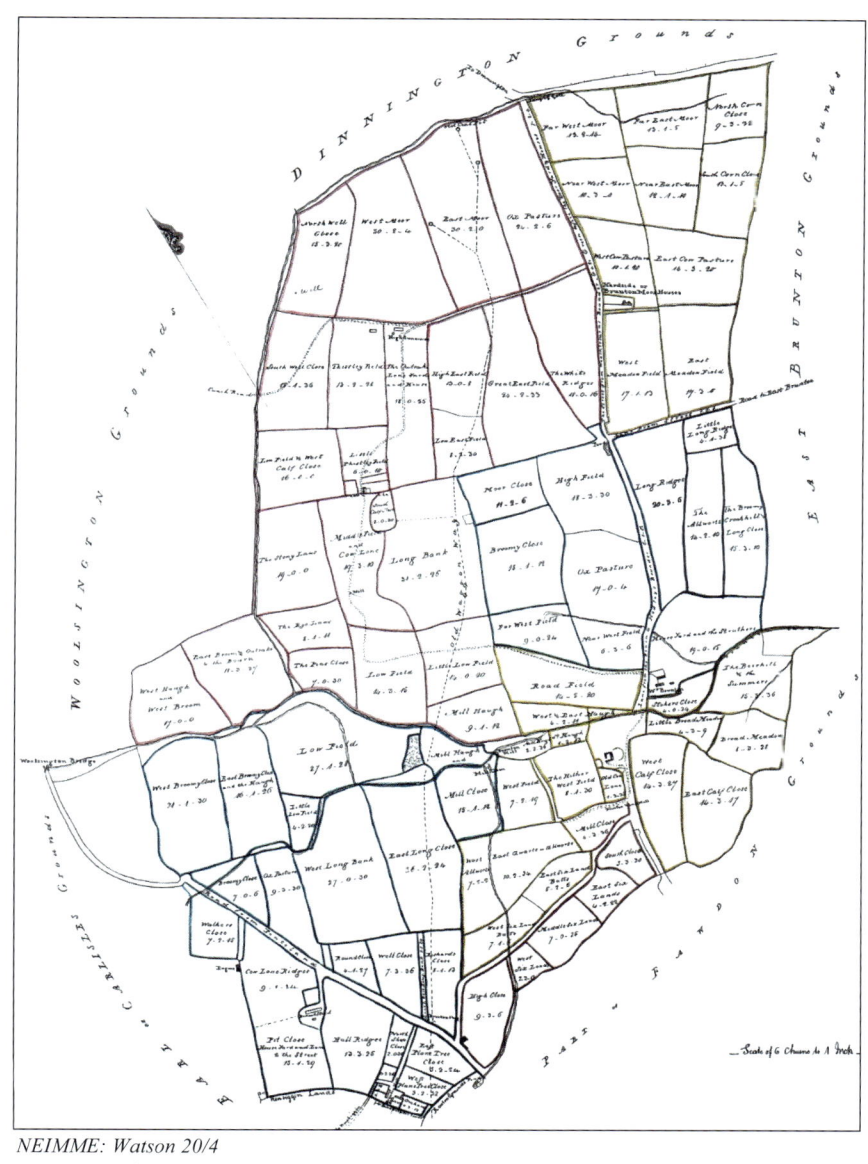

Fig. 4: West Brunton Waggonway to Scotswood

---

[12] NEIMME: Bell 14/303, 305.
[13] NEIMME: Watson 20-4.

To the west of Brunton, Richard Peck and Mr. Beaumont senior had been mining in Whorlton Moor and Newbiggin since at least 1738 using wains to transport their coal to Lemington. However, in 1752, their sons, Robert Beaumont and William Peck, planned a waggonway from Newbiggin, through the manor of Newburn, to staiths on John Roger's land at Lemington. The development of Holiwell Reins Colliery was expensive: in 1752, William Peck recorded that 'the Winning of the Colliery, laying the Waggon Ways, Building the Steath etc a very great expense – suppose £10,000'. His estimate of the running costs illustrates the high cost of transport which amounted to 42.4% of the total budget because the waggonway, unlike the wainway, took a circuitous route to avoid climbing Coley Hill.[14] The line was completed in 1756 and the waggonmen were paid 1s 4d for a return trip of about eight miles. A normal day's work would have probably been three trips enabling the men to earn 4s a day which amounted to a potential annual income of about £55. The cost of keeping the horse would be approximately half that sum. This still left a tidy income of about £27 for the waggonman which was roughly equivalent to the earning of a good hewer, the best paid workers in the mine.

The plan of Holiwell Reins Colliery (fig.5) shows pillar and stall workings in the High Main seam which was the standard method of mining coal in the northern coalfield at this time. On February 27th 1767, Christopher Bedlington recorded in his view book that 'there are four Engines kept going at this Colly. and has never got the water out since last flood'.[15] The four engines are marked on fig.5 as is the network of waggonways leading from the individual pits. This is a good example of the early type of colliery railway which served numerous shallow pits spaced about two hundred yards apart. The pits lasted for about six years after which a new pit was sunk and the waggonway moved. Consequently, there were many branches occupying almost every field in the area of mining which was in effect a large marshalling yard. These branches came to a head at the boundary of the royalty. The stem of the waggonway, which was a much more permanent feature, proceeded through the Duke of Northumberland's land for four miles to the staith at Lemington. When deeper shafts became necessary, the pits were fewer in number and spaced about a mile apart. Then the railways to marshal the coal were built underground.

Holiwell Reins became a very successful colliery. However, in the late 1770s, production fell from 10,432 chaldrons in 1778 to 4,989 chaldrons in 1780. On September 7th 1780, a notice appeared in the local press advertising the sale of 'Holywellreens Colliery, Several Gins, Waggon HORSES and underground GALLOWAYS all in good Condition. Also a Number of Waggons, nineteen Bolls each, two Coal Gins and several other Colliery Materials. Also about four Miles of old Waggon Way to be sold as it lies'. In addition, there were two 'fire engines' for sale. All enquiries were to be made to Christopher Bedlington at Walbottle. The Newcastle businessman William Cramlington bought Holiwell Colliery which was allocated a share in the seasale trade from the river of 14,000 chaldrons in 1788; but by the beginning of the nineteenth century it is not listed amongst the seasale collieries.[16]

---

[14] NEIMME: Forster 1/5/154 and 155
[15] NEIMME: East 10b; map Watson 24/29.
[16] NEIMME: Bell 17 p. 299. Watson 2/11/2 and 110.

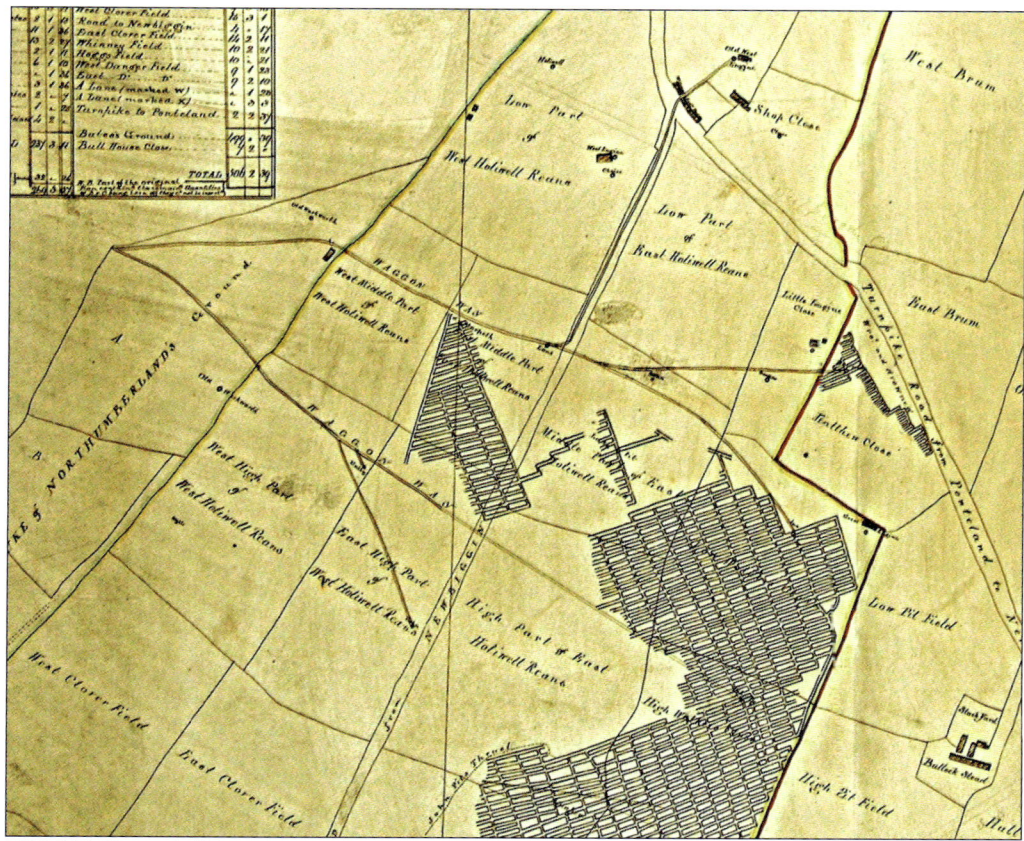

NEIMME: Watson 24/29

Fig. 5: Holiwell Reins Colliery circa 1767

The Holiwell Way, which was opened in 1756, was not the first waggonway to Lemington: that distinction probably belongs to Newburn Moor Colliery Way which is shown on Isaac Thompson's map of Denton estate in 1754 (Fig.6). This colliery was leased from the Earl of Northumberland by Mr John Humble, a member of a prominent family in the coal trade who operated other collieries in Ryton, Shiremoor and Birtley. Newburn Colliery was equipped with a Newcomen pumping engine which was built by William Newton and his apprentice John Watson. William Brown was the viewer there.[17] Four pits are marked, one of which is called the Turnrail Pit, suggesting that a short waggonway was used to carry the coals downhill to Humble's staith. The road descending the hill from the north is the old wain route from Whorlton Colliery. When Isaac Thompson made his survey of John Roger's estate in Denton, Humble's colliery was nearing closure and West Denton Colliery had been closed temporarily.

Isaac Thompson's map also shows a second staith further south on the Earl of Northumberland's land at Lemington, which was the terminus of William Brown's waggonway from Throckley, opened in 1751.[18] Initially, the Throckley Way was built as a main way only, but by the autumn of 1754 a second line, the byeway, had been added which caused some distress to Widow Bell, the tenant of Newburn Hall, who claimed compensation for damages to her grazing lands. The reason for the addition of the byeway was probably to accommodate the extra traffic from Heddon Colliery which

---

[17] NEIMME: Watson 23/16; NRO: Sant/Beq/09/01/01/19; Watson 2/4/ 14,29,56 and 120.

[18] Sir Hugh Smithson succeeded as Earl of Northumberland in 1750 and was created first Duke in 1766 which can cause some confusion in reading documents of this period.

had been leased by the Throckley partnership in 1753. The correspondence between Henry Masterman and William Brown reveal that greater care was taken to make the main way level 'there being no necessity to have a Byeway to run so much upon a Horizontal'. Brown resisted widow Bell's claim for damages on the grounds that 'it is commonly understood that when a Waggonway is taken that their must be both a Main Way and Byeway'. However, he was overruled by Masterman.[19]

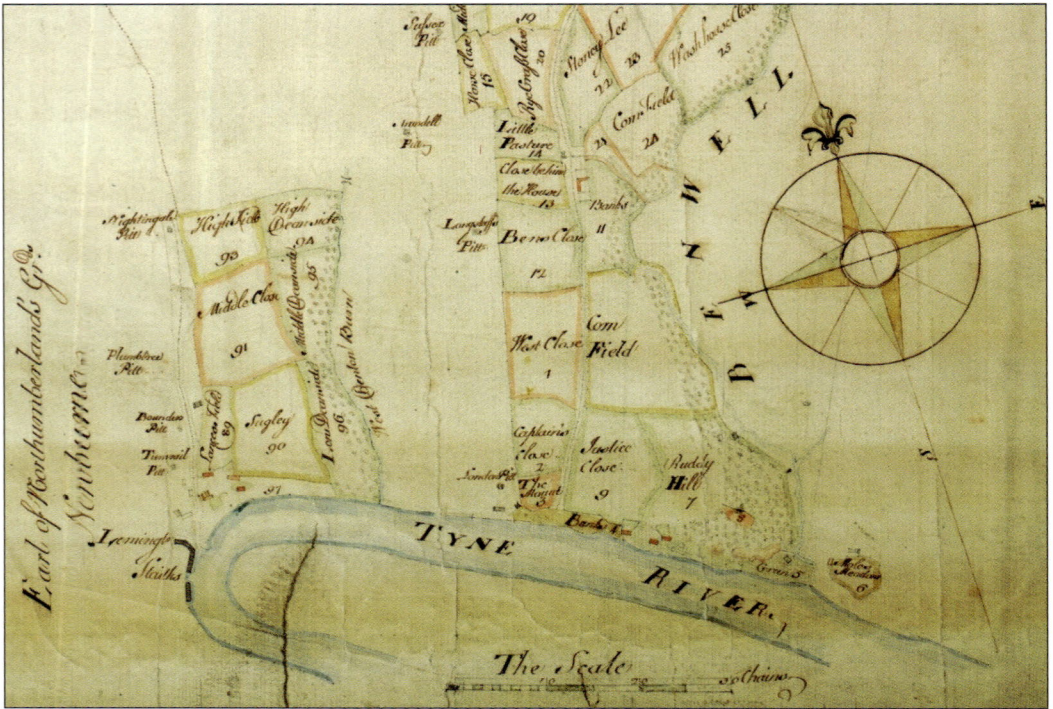

NEIMME: Watson 23A/16

Fig. 6: Part of Isaac Thompson's Plan of Denton 1754

The bend in the River Tyne and the wain road marked the boundary between the Earl of Northumberland's land in the west and John Rogers' Lemington estate in the east. In 1756 two more staiths were built at Lemington, not on the Earl's property but on the part of John Rogers' land marked as 97 on Thompson's plan.[20] The western staith, 90 yards long and 22 yards wide with provision for six keels, served Robert Beaumont's colliery at Holiwell Reins; and the eastern staith, with provision for four keelrooms, served John Blackett's colliery at Wylam five miles to the west.[21] Later a fifth staith was added for John Rogers' Lemington Colliery which had provision for two keels. After Edward Montagu had inherited John Roger's estate in 1758, this landsale colliery was elevated to a seasale colliery to provide finance for Montagu's main enterprise in East Denton, which became known as 'The Monty'.[22] By 1756 Lemington was served

---

[19] NEIMME; Brown/1/302, William Brown to Henry Masterman 30th November 1754.

[20] John Rogers suffered a mental illness and was declared a lunatic in 1744. His interests were administered by Charles Montagu who inherited East Denton estate which included Sugley in 1758. William Newton was his viewer.

[21] NEO: Sant/Gen/1/4/2/4. The location of the staiths would suggest that the Holywell line was built first. After leaving the Throckley Way, the Wylam Way probably ran beneath the trestle bridge which carried Holiwell coal to its staith. Blackett decision to build on John Roger's land rather than his own in West Denton was doubtless determined by the difficulty and expense of crossing Sugley Dene.

[22] NEIMME: East 10A p.74 Bedlington records that on 28th April 1766 'made measurements of West Denton staith at East end of Mr. Blackett's at Lemington' which was two keel births long (30 yards) and 22 yards wide.

with three major waggonways: the line from Throckley and Heddon collieries; the line from Holiwell Reins to the north and the one from Wylam to the west. When the pumping engines were finally installed in the southern part of Throckley estate in about 1760, a branch was built to join the Wylam line at Throckley Reins. Isaac Thompson's plan of Newburn in 1767 (fig.7) captured a waggonway system on the eve of major changes brought about by the building of the fourth engine for Throckley Colliery and the development of Walbottle Colliery. This is discussed in chapter five.

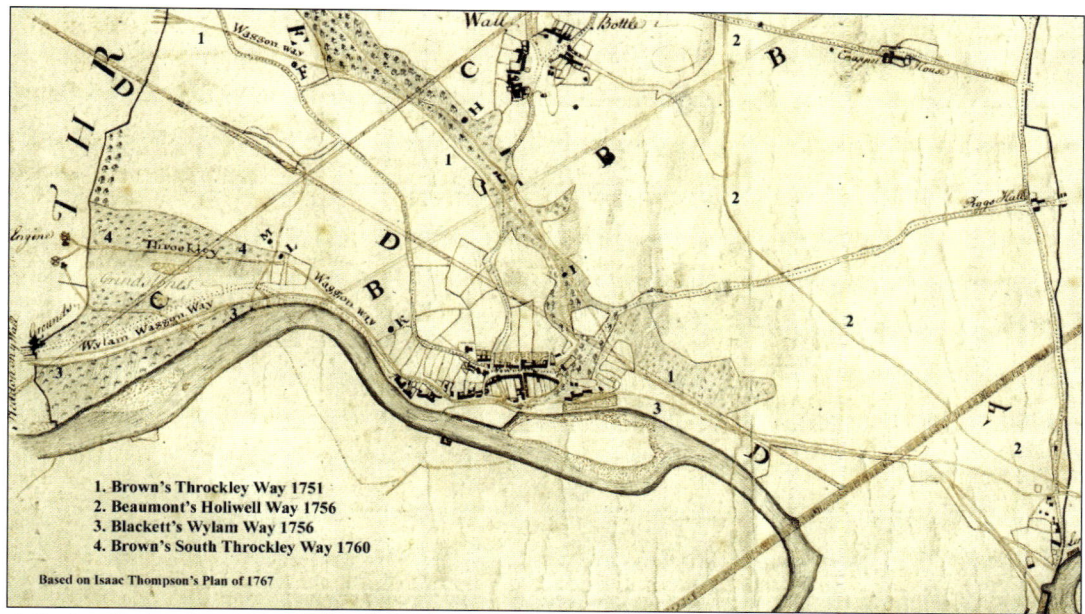

1. Brown's Throckley Way 1751
2. Beaumont's Holiwell Way 1756
3. Blackett's Wylam Way 1756
4. Brown's South Throckley Way 1760

Based on Isaac Thompson's Plan of 1767

Author's collection: based upon NRO:SANT-BEQ-05-03-14-36

Fig. 7: Major Waggonway Routes to Lemington in 1767

Wylam Colliery to the west was an outlier separated from Heddon by a forty fathom downcast dyke near the boundary of Colonel Bewick's Close House estate, which was barren of coal. The High Main seam was only six fathoms from the surface and had been worked since medieval times; and there were four other worthwhile seams less than forty fathoms below. Although Wylam Colliery achieved notoriety in the early nineteenth century, when William Hedley was experimenting with iron plate rails and then steam locomotives to cut the cost of horse transport, Wylam was the least significant colliery in the area of the golden triangle during the eighteenth century. The waggonway, which followed a largely level route parallel to the river for five miles, had no significant engineering features to compare with the wooden bridge across Walbottle Dene on the Throckley Way or the great cut and embankment at Cutend on the extension of the railway northwards to Callerton. However, a small stone bridge across the New Burn survives to this day. Also Wylam Waggonway had an insecure existence in its early years. During his negotiations with Edward Montagu in 1762, John Blackett threatened to build a canal to replace the railway. Doubtless he was influenced by the success of the pioneering work of James Brindley in building the Bridgewater Canal from the Duke's pits at Worsley to Manchester, which had opened the previous year. Montagu's colleague, William Archdeacon, reminded Edward that 'if Mr Blackett makes a Canal as it is affirmed he will in conjunction with others' revenue will be lost. The event certainly illustrates that in the mid eighteenth century John Blackett of Wylam was at the forefront of developments in transport as Christopher was later.[23]

---

[23] NRO: Sant/Gen/1/4/2/8. Sant/Gen/1/4/2/10.

John Blackett also owned West Denton until it was sold to John Baker in the 1760's. A large geological fault, the Great Dyke, bisected Denton estate and threw the strata to the north ninety fathoms down which had a major impact upon mining: while the important High Main seam outcropped in the northern part of the estate, it was absent from the southern part, where the Low Main was the seam nearest the surface. It was this easily accessible Low Main coal near to the river which was mined for the medieval, Elizabethan and Jacobean coal trade. In about 1748 Beaumont leased West Denton Colliery and installed a Newcomen engine to win the coal seam thirty fathoms below which was named the Beaumont after him.

Edward Montagu's achievement was to win the Beaumont seam in East Denton and open up the Montagu Colliery in 1766. In this task he encountered problems with his neighbour, John Blackett, and Montagu was at the point where proceedings at the Court of Chancery were threatened; but this was a course of action which his viewers, William Brown and John Wright, thought inadvisable. Blackett's colliery, which had been abandoned in 1758, was to the rise of East Denton and the flooded waste threatened the development of both East Denton and Benwell beyond. Edward's advisers pointed out that 'without the Engines…in West Denton your Honour's Colliery…will labour for many years under the inconvenience of wet Coals and we shall lie constantly under the apprehensions of being irrecoverably drowned out by Mr Blackett who has it in his power to turn the River Tyne or West Denton Burn into his colliery all of which water would unavoidably fall upon East Denton Engines and over power them'.[24] That such action was contemplated is an interesting comment upon the business tactics of coalowners at the time. Edward Montagu was advised to abandon his plans for legal action and 'to procure leave of Mr. Blackett for to put in proper repair one of his Engines at West Denton' since 'if these drowned wastes were drained the Colliery at Lemington might be set to work very probably in 6 or 7 months which otherwise could be in so many years'. However, 'if a Chancery Bill is to be preferred it…would be productive of such ill consequences as might render the winning (of Lemington and East Denton) utterly impossible'.[25] The matter seems to have been resolved by John Baker's purchase of West Denton estate – the Baker family having been associated with the Montagu family since the beginning of the century.

By 1773, the miners at East Denton Colliery were working both the Low Main and the Beaumont seams further north near the Great Dyke. Beyond the Great Dyke, the High Main coal was present twenty three fathoms below the surface. This high quality household coal, and the much inferior Newbiggin Stone Coal, had been worked by Rogers and Carr before Edward Montagu inherited the estate in 1758. In the next decade, the workings of the High Main seam and the Newbiggin Stone Coal north of the Great Dyke were extended. After Elizabeth Montagu's purchase of West Kenton in 1779, the colliery continued its progress northwards and by 1796 the High Main coal in West Kenton was being mined. These pits were served by a waggonway running down the west side of Denton Burn to the staith near its confluence with the River Tyne.

To the east, Benwell and Elswick estates occupied a steep escarpment which fell almost 400' in less than a mile from the top of Westgate Hill to the banks of the River Tyne. The High Main seam outcropped roughly on the line now followed by Elswick Road and had been mined since Roman times. Near the river, the Low Main seam was only

---

[24] NEIMME: Watson 2/8/102. See also Watson 23A for plans of the workings
[25] NEIMME: Watson 2/8/102. My parenthesis.

twenty feet beneath the surface and it had been mined since medieval times when both of these manors belonged to the priors of Tynemouth. After the dissolution of the monasteries, they came into the possession of the king and a valuation of the Crown's mines in 1611 recorded that there were two seacole collieries in Benwell, rented for £20, which produced a profit of £120 – a very handsome return. One of the earliest waggonways in the north is recorded on a map of Benwell dated 1637.[26] The colliery was drowned in about 1670. However, in 1698, Charles Montagu and George Baker took out a lease for 31 years and Benwell once again became a very important colliery before flooding forced the partnership to abandon the mine. Montagu probably retained the lease to prevent it falling into the hands of his competitors. The colliery was rewon in 1766 by means of a massive steam engine, with 75 inch cylinders, built by William Brown. Christopher Bedlington was his assistant responsible for the work and his view book contains a detailed description of the progress of this work. Elswick estate was also part of Montagu's coal empire in the early eighteenth century and one of the first Newcomen pumping engines on Tyneside was erected near the river in 1722. However, the colliery was drowned by 1740.

NEIMME: Dunn
Fig. 8: An Eighteenth Century Coal Pit

The early history of the large estate of Coxlodge on the northern side of the Town Moor is sketchy. In 1759 William Brown provided an estimate to a Mr Collingwood for a steam engine to drain the colliery; and in 1776 he surveyed the route for a waggonway which followed the edge of the Town Moor then turned south through Jesmond to the mouth of the Ouseburn. A branch line was planned from Sandyford to Barras Bridge to capture the landsale market in Newcastle. There is no firm evidence that this route was built but the amount of coal extracted from the royalty would suggest that it was: according to a valuation of Coxlodge Colliery in 1808, half a million tons of the best quality household coal from the High Main seam had been mined by that date. The coal was extracted from the 105 acres to the south of the Great Dyke and the transport problems associated with moving such a large tonnage would suggest that a waggonway served the colliery.

The southern section of Brown's proposed waggonway from Coxlodge, the section from Sandyford Bridge (Lambert's Leap) to the River Tyne, was almost certainly using the old Jesmond Way which was probably built by Richard Peck the leasee of Matthew White's colliery. The line had been abandoned in 1745 when the colliery was flooded. The geography of Jesmond placed great restrictions on the choice of the route for a waggonway since the descent to the Ouseburn was precipitous. The route from Sandyford Bridge is a controlled descent down the western side of the valley. This was the most practical option for a railway and it is still used by today's roads – Stoddart Street, Stepney Road, Lime Street and Ouse Street.

---

[26] The map is available in History of Northumberland, Vol.XIII p.230, and has a claim to be the oldest railway map in the world.

Such was the history of Brown's homeland. The collieries of the golden triangle in the days of William Brown were major enterprises using the lastest technology like those elsewhere in the Great Northern Coalfield. Besides supplying the local landsale trade, the principal part of their business was selling domestic fuel to London and the South East of England where most of their profit was made. The expansion of this trade was made possible by building an extensive network of waggonways and several were William Brown's handiwork. Most of these early railways, like those elsewhere in the Great Northern Coalfield, were complex structures not simple tram roads. The bridges, cuttings, embankments and staiths along their routes were regarded as major civil engineering achievements in the eighteenth century. These private railways were built and operated by a skilled workforce. Some lines were short lived but others, like the Wylam Way built in 1756, served the industry for more than a hundred years; and afterwards the route survived as part of the public railway network for another century. Steam engines, in the form of Newcomen pumping engines, were used extensively to drain the mines within the golden triangle. The mines of Brown's homeland were some of the first collieries to use this new technology beginning with Charles Montagu's enterprise at Elswick in the 1720's and Richard Peck's work at Jesmond in the 1730's. More powerful engines with larger cylinders were built at Throckley, Walbottle, East Denton and Benwell by William Brown and his team in the 1760's. Thus, there was an extensive pool of engineering experience within the golden triangle before the days of Hedley, Hackworth and Stephenson. Indeed, these men were a product of a long engineering tradition established by colliery viewers such as Richard Peck, Christopher Bedlington and William Brown. During the eighteenth century, one hundred years before George Stephenson built the Liverpool to Manchester line, steam engines and railways were a familiar sight in what today is known as the western suburbs of Newcastle upon Tyne. These stationary engines and wooden waggonways were the forerunners of the famous Wylam Railway: Hedley's travelling engines and iron railway did not mark the beginning of the railway age but the culmination of a century of technical development in which William Brown played a major part.

Fig. 9: Hedley's Wylam Way – the Culmination of a Century of Development

Chapter Two - The William Browns of Throckley Pitt House

It was the custom in the eighteenth century for the eldest son to be named after his father which is a source of much irritation amongst today's historians. William Brown, the subject of this book, was born at Heddon Pit House in 1717 where he lived with his father and grandfather – both called William Brown.[27] In 1711, there was a William Brown working on James Clavering's waggonway from Chopwell to Stella. In the same year William Cotesworth's accounts note 'Cash pd Wm Browne for 5 days Loading Railes into ye Way with his draught ass... 12s 6d'. The possession of an ass suggests that this man was no ordinary labourer. The accounts also record payments to Nath. Browne and James Browne for laying the waggonway.[28] It is tempting to link this family of railway workers with our man but both Brown and William are common names and the link to the Brown family living at Heddon Pit House has not been firmly established.

William Brown's father was married to Ann Watson, the daughter of Lewis and Jane Watson of Throckley Pit House. Lewis died in 1732 and this was probably the occasion for the Brown family moving from Heddon to Throckley where they lived for the next forty odd years. William Brown's father is listed as the tenant of Throckley Colliery in an account of the rentals of the late Earl of Derwentwater's estate in 1735; and this is undoubtedly the same man, William Brown of Throckley Pitt House, who is named in three indentures, dated 1737.[29] His landsale colliery in Throckley appears to have been in poor shape for he is cited as owing £82½ to Greenwich Hospital, more than three times the yearly rent of £25. The documents record him leasing Newbiggin Colliery (with the exception of those areas leased to Richard Peck) from the Earl of Carlisle. This colliery was then assigned to the hostman Christopher Rutter to whom both William Brown and his son were bound on a penalty of £100. Like Throckley, Newbiggin was primarily a landsale colliery supplying the neighbourhood with house coals at six shillings a chaldron and inferior coals for use in brick and lime kilns at 4s 6d a chaldron, the same rates that Richard Peck was charging at Whorlton Colliery. However, there was an agreement whereby Brown also supplied twenty keel loads of seacole (defined as 160 chaldrons) to Christopher Rutter for which he received eight shillings a chaldron; and the revenue from the sale was to be used to pay off Brown's debt to Greenwich Hospital.

William Brown's neighbour was the viewer and agent of the Earl of Carlisle, Richard Peck of Newbiggin, one of the earl's estates which is now part of the western suburbs of Newcastle. Peck held a lease of Jesmond Colliery, to the north east of Newcastle, from Matthew White. Peck was a close associate of Amos Barnes, the Grand Allies viewer at Heaton Banks Colliery. By virtue of their skill as builders of Newcomen engines, these men established the greatest concentration of steam power in the world in the lower Ouseburn valley during the 1730's. In October 1732, Peck recorded that 'the 2nd Coale Pitt in the Low Ground Called the Nicholas sunk by Brown and Co'. Two years later he noted that 'the Sinking Pitt in the Stone Style Close near Thomas Naters Stockgarth Called the Chance Sunk by Robert Brown and Co'. Peck's records also contain a report upon Jesmond Colliery made in 1739/40 by ten viewers one of whom

---

[27] See Appendix II.
[28] TWAMS: Cotesworth Paper CK/11/64, 65, 67.
[29] Morrison J. Newburn Manor p.144; NEIMME: Les/3/54, 55 and 56. Brown's letters record that in April 1752 he was in London negotiating a new lease for the customary period of 21 years with the Commissioners of Greenwich Hospital. Assuming that the previous lease was a similar length, the would indicate that his father took a lease in 1731.

was Jonathan Brown. Once more the link with the Brown's of Throckley is tantalizingly close but not confirmed.[30]

The historian of Heddon, Cadwallader Bates, records that in the early 1730's, Heddon Colliery was leased by the Earl of Carlisle to a Mr. Barcus who employed William Brown as his overman.[31] This comment undoubtedly refers to the father since his son, born in 1717, was only a youth at that time. Bates then notes that:

> 'Brown was a remarkably able man, and when afterwards Mr. Barkas threw up his lease owing to the bad state of trade, the story goes that in buying some flannel for his pit clothes from Mr. Bell, a wealthy draper in Newcastle, he happened to mention what a pity it was that the Heddon pits should be laid in, and the partnership of Bell and Brown was consequently formed to work them, and the adjacent royalty of Throckley'.

Again, these comments refer to the father, not the son as has been generally believed. The Bell family certainly ran a prosperous drapery business in Newcastle but they were no strangers to the coal trade. Matthew Bell is mentioned in the Vend for 1728 as the owner of Brockwell Colliery in Winlaton which transported its coals by a waggonway northwards to Bell's staiths at Stella. In 1735, Matthew Bell was admitted to the powerful Hostman's Company. He now had both the money and the credentials to engage more fully in the coal trade which Brown, being in financial difficulties with Throckley Colliery, was probably keen to exploit. This was the beginning of partnership between the two families which was to lead to their ownership of more important collieries at Shiremoor, Willington and Bigges Main later in the century.

William Brown's cousin was John Watson (1729 – 1797) one of the apprentices of William Newton of Burnopfield, the distinguished viewer of the great collieries of the Lambtons and Windsors. It may be that William Brown had served his apprenticeship with William Newton in the 1730s for they were related through their connection to the Watsons. The other possible candidates are his father or his neighbour Richard Peck. Unfortunately, there is a gap in our knowledge between 1737, the date of these leases for Newbiggin Colliery, and 1749, the date of the earliest surviving letters of William Brown. These were written to Carlisle Spedding of Whitehaven, the viewer of Lord Lowther's collieries in Cumbria. In 1713, Carlisle Spedding had been sent surreptitiously across to Tyneside, the centre of mining expertise, to learn the advanced skills of mining and it is tempting to speculate that William Brown's father and grandfather began the family's friendship with the Speddings at that time.[32] This gap in our knowledge between 1737 and 1749 is unfortunate for these were clearly formative years for William Brown who was gaining a reputation for his expertise in designing waggonways. Carlisle Spedding sought his advice for his projects in Cumbria and there was an interesting exchange of letters concerning waggonways to the Whitehaven pits. In 1754, the Duke of Hamilton commissioned him to lay a waggonway underground at Bo'ness Colliery near Edinburgh. However, it is not clear where, William Brown gained this experience in waggonway construction to merit his reputation. The obvious place to learn his trade was the network of waggonways leading

---

[30] NEIMME Forster 1/5/54, 1/5/57, 1/5/112, 1/5/120.
[31] Archaeologia Aeliana, New Series XI p.285.
[32] Carlisle Spedding (1695 – 1755) was the youngest of four sons. He worked in Sir James Lowther's mines near Whitehaven in Cumbria where his eldest brother was the agent.

to the staiths at Stella on the southern bank of the River Tyne in sight of Throckley Pitt House. Perhaps he was a part of the family business referred to in the Cotesworth papers and in Peck's view book.

There is also a lack of detail about William Brown's personal life. In 1741, he married Mary Smith of Morpeth by whom he had three sons and four daughters. Tragedy struck the family in 1748. A plaque on the outside wall of Heddon Church records the death of his youngest son John, aged three, on 19th January; and two weeks later the death of his infant daughter Agnes, aged ten months. This tragedy serves as a reminder of the high rate of infant mortality in an age before adequate medicine was available even to the better off families. William Brown appears to have had a genuine affection for his children: in 1765, he wrote movingly to a business associate, Captain Snowden, that 'Poor Bill my son is lying very ill in a fever and am sorry to say is not out of danger'.[33] His son and heir, did survive and, in 1770, he married Margaret Dixon of Hawkwell in Northumberland. A third son, Richard, trained as a viewer and became part owner of Washington and Wylam collieries. Interestingly, there are hints that the family may have been connected with the famous landscape gardener, Lancelot – 'Capability' – Brown of Kirkharle (1716 – 1783).[34] Besides being contemparies from the same area, they had common business associates in men like the Earl of Carlisle, the Duke of Northumberland and the Duke of Portland. William travelled with Lancelot's brother John Brown of Kirkharle, the agent of the Duke of Portland, to inspect Hucknall Colliery in Nottinghamshire and Warnell Fell Colliery in Cumberland. William and John were also in correspondence about the running of Black Close Colliery near Ashington one of Portland's properties. Certainly, the families knew one another but whether they were related has yet to be proved.

Like many gentlemen of the Georgian period, William Brown had an interest in antiquities and is credited with saving some statues from his local Roman fort at Rudchester on Hadrian's Wall. However, he was also engaged in removing antiquities from the region for private collectors. In August 1754, he received a letter from Francis Woodhouse, a friend in London, who wrote 'at Carvoran bought a stone which cost me 2s and 1s paid for the carrier to carry it to you….one Thomas Harden, a poor man of that place, has promised to get some more which I desire you'll pay for, but not to exceed 2 or 3s each and the Inscriptions are to be fair'. In October, he notified Woodhouse that 'I have Captn William Fall of the 'Rose in June' putt on Board the Stone you sent out of the West Country and also a brace of moor Game and a couple of Dryed Salmon which please Except of which I hope come safe and sound'. Clearly, there were other riches in the region besides coal.[35]

The letters reveal that William Brown was a man of many talents. His work was not confined to the Great Northern Coalfield for much of his early correspondence refers to the copper mines at Middleton Tyas, near Richmond, in North Yorkshire and the lead mines at Grassington and Swaledale. Some claim that William Brown was the first to introduce the screening of coals; Galloway records Brown's attempts to mechanise the hewing process with a machine popularly known as 'Willy Brown's Iron Man'; he was

---

[33] NEIMME: Brown/2/William Brown to Captain Snowden, 11th February 1765.
[34] Alnwick Castle: DP/D1/1/124 Richard Brown at Kirkharle in connection with the Ponteland Turnpike road, 16th September 1777. Lancelot Capability Brown's father was another William Brown. Nottingham University, Duke of Portland Archive PwF 1735/1746.
[35] NEIMME: Brown/I/231 Francis Woodhouse to William Brown, 22nd August 1754 and p.232 William Brown to Francis Woodhouse, 2nd October 1754.

involved with Michael Menzies in introducing an improved method of hauling coal within the pit[36]; and he was senior viewer at Hartley when the Delaval's agent, Joseph Oxley, introduced steam winding. His early reputation as a builder of waggonways has already been noted; and later in his career he was to design the important network of lines in the Tyne Basin, the line from Harraton to the Wear, the line serving Washington Colliery which led to both the Tyne and Wear, and Morton Davison's waggonway from Beamish to Fatfield, the rival to the Grand Allies Tanfield Way.

His talents were diverse but he was primarily recognised as the builder of steam engines: as Matthias Dunn, one of the first government inspectors of mines and an early writer on the history of the coalfield, observed, 'Mr William Brown of Throckley ... was remarkably conspicuous in the introduction of the steam engine to this colliery district'.[37] The most comprehensive account of William Brown's work as an engine builder is provided by Sam Haggeston, the 70 year old furnace keeper at Hebburn Colliery. In 1811, Matthias Dunn, who was the underviewer at Hebburn, collected Sam's recollections of his life in the mining industry which later formed the basis of Dunn's 'History of the Viewers'. Sam recalled that, about the year 1750, William Brown 'began to turn his attention very much to the improvement and application of the Steam Engine which was then in a very rude state'. He then named twenty three collieries where Brown had built a total of twenty nine engines.[38] This is an impressive record. Brown was responsible for the introduction of large iron cylinders from Coalbrookdale (60 to 75 inches in size compared to the 42 inch cylinders in use in 1750) which made more powerful engines possible. He appreciated the need for a plentiful supply of steam and built multiple boilers to improve the supply. He also used larger pipes – the 24 inch wooden pipes for Benwell being three times the size normally used. Sam commented that, after winning Benwell, Brown 'became very famous for his Knowledge of Engines'. His work with Newcomen engines is discussed more fully later in chapter seven dealing with his greatest achievement – the conquest of the Tyne Basin which necessitated power pumping engines sinking shafts over one hundred fathoms deep to the valuable High Main coal.

William Brown's correspondence provides a fascinating insight into both the business world and the mining industry of the eighteenth century. In a letter to Leonard Hartley, the estate manager at Middleton Tyas, he commented that 'I am ready and willing to Serve you in everything in my power nor do I value time nor Distance of Ground to Serve you therefore Command me'. This appears to have been his core business ethic: his service consisted of supplying advice, equipment and personnel to clients throughout the country. In January 1755, William Brown wrote a follow up letter to John Burrell, the leasee of the Duke of Hamiliton's colliery at Bo'ness, where he had built an underground waggonway in 1754. The letter illustrates the service which

---

[36] Menzies is credited with introducing the inclined plane underground.
[37] Taylor, Archaeology of the Coal Trade, p. 195 records 'the practice of separating small coals by screening, about the year 1740, by Mr. W. Brown. There is some confusion about the date which is sometimes given as 1760 and attributed to Willington Colliery; but Brown did not take out the lease of Willington until 1773 and coal was not raised until November 1775. Galloway, R., 'Annals of Coal Mining and the Coal Trade' (London 1898) p.305. NEIMME: Transactions XV p.209 Chronicles of the Coal Trade. Dunn, M., 'An Historical, Geographical and Descriptive View of the Coal Trade of the North of England' (Newcastle upon Tyne 1844) p.41. Several of the dates given by Dunn, which were later repeated by Galloway, are not correct.
[38] NEIMME: East 3b. This manuscript is invaluable because of the scarcity of other information on the subject. The collieries were Throckley, Birtley North, Lambton (2), Byker, Walker (2), Bell's Close, Heworth (2), Shiremoor (2), Hartley, Walbottle (2), Oxclose, Beamish, Benwell, West Auckland, North Biddick, Pittenweem, Bowness, Muselburgh, Low Fell, Fatfield, Washington, Willington (2) and Felling. See Appendix I.

Brown offered to his clients and also demonstrates that the North East was already exporting its technical expertise in railway building outside the region:

> 'pray how goes the new way forward underground and what steps have you taken with the intended Waggon Way above ground. Will be glad to have these designs going forward. If want a man to Build Waggons or Lay a Waggon Way can serve you a very good one that will willingly come to your part'.[39]

It was during his visit to Bo'ness in 1754 that William Brown had been disturbed by the employment of women underground, a practice which did not take place in the Great Northern Coalfield. In a letter to Leonard Hartley he expressed his misgivings:

> 'at Burroughstoness I was assisting to Contrive a Waggon way underground on which will have to bring coals 600 yards....at present they are conveyed from the face of the workings to the pitts Bottoms and in some places up the shafts upon womans Backs which poor Creatures Carries some of them 13 Stone weight of Coales at a time in these Creals'.

To which Hartley replied:

> 'what could the owners of that Colly been thinking of that before could never contrive some means to convey the coales otherwise than be making these poor Women so Wretched'.[40]

Their reaction predates by almost a century the popular outcry to the same revelation by the famous 1842 Children's Employment Commission. William Brown's sensitivity to the suffering of others is revealed elsewhere in his correspondence. In April 1765, he described the accident at Walker Colliery where 'poor Renwick and another man was terribly burned', eight men and fourteen horses killed and 'the dismal scene the poor wifes loss makes is not to be expressed'.[41]

Not only did he supply men capable of building waggons and waggonways, he ran an agency which supplied other craftmen to the mining industry such as borers, engine wrights and blacksmiths. Writing to Edward Harrison of Kendal regarding hiring a borer, he commented that 'I have prevailed upon one of the Best in our Country to consent to come to you; he is of unblemished character and has good Experience in that way as any in our Country has Rodds and all untensells of his own so you will have little trouble with him or the carrying on of that branch of business'. Earlier he had written to Carlisle Spedding that 'The deepest (pit) we have in our Country is 83 fath. It was Bored by one Jno. Rawling who is famous for that business in our Neighbourhood and has no difficulty in boring Deeper but charges very dear for his work. Tho he gives the Truest Acco't of the Strata he bores Through that Ever met with. Our Chizells are Commonly 2½ Inches unless in Extraordinary Cases in which have known them 4 inches.'[42]

---

[39] NEIMME: Brown/I/240 William Brown to John Burrell, 20th January 1755.
[40] NEIMME: Brown/I/182 William Brown – Leonard Hartley 30th Nov. 1754; p.185 Leonard Hartley – William Brown 9th Dec. 1754. Hartley was a land agent probably for the Middletons. The family owned land in the village and his son had chambers in Lincolns Inn.
[41] NEIMME: Brown/II/33 William Brown to George Douglas 4th April 1765.
[42] NEIMME: Brown/I/175 William Brown to Leonard Hartley 25th September 1754; p. 108 William Brown to Edward Harrison 14th Oct. 1752; p. 14 William Brown to Carlisle Spedding 30th Oct. 1750.

There is a fascinating set of letters regarding the fitting out of a lead mine at Low Row, upper Swaledale, in which Leonard Hartley asks 'pray be so good as talk with Ord the Blacksmith and ask him If there be any prospect of getting a Young Man of his Trade we shall want one to attend our Mine here. His chief Business will be to Sharp Tools.' The pay was to be 8s a week 'but of all things he must be Sober'. Brown replied 'as to the Smith I have promised one that I am Sure will please. He was Ord's man when most of your things were made and Indeed made most of them. His Master and he Differed some time ago about Wages and he has since been working at a new fire Engine that is Erecting in our Country and being willing to take your wages waits on…..orders from you to Set forward. He has a wife but no Children. Let me know if have any objection to his being Married for, if has, can have a journeyman that will come.' Hartley did object replying that 'a single man is more Desirable If he is as Sober and well Disposed'. He also asked Brown to supply the equipment for the smithy: 'pray be so good as buy or bespeak a stoody (anvil) of about Ten Stoneweight, a pare of Bellows Suitable to Such Bussiness as ours will be which you or your Workmen Ord can Judge Very well, a Common working hammer and Striking Hammers. As soon as they are Ready send them by the first ships to Stockton to the sure of Mr James Lambe Wine Merchant there who shall have orders to pay the baggage and as soon as I know the charge shall…pay the Money'. He also orders 'a small quantity of Nails of Several Sorts that may be useful in the Shafts and other works'.[43]

In the discussion regarding pumping machinery, a glimpse of Brown the inventor is revealed: 'there is a much Simpler Machine that I have contrived to work with watter also that if the Situation does but Suit will not cost ¼ of the Expense. It has neither wheel nor Bobb nor Crank. I have it now working Underground at Throckley with great Success. It has been seen by most of our knowing people who all agree it is the Simplest and best thing of the kind they have seen'. Thos. Parke one of the partners commented that 'I have shewd the Plan of your Machine to my father and partners who very much approve there of and think themselves greatly Indebted to your Mechanical Genieus.' However, Brown found that there were problems with his new equipment: 'I have just now Discovered an Imperfection in it that I wd not Recommend it to any friend till I contrive ways and means to Remove the Imperfection.' This was a remarkable example of his honesty upon which so much of his business depended.

Brown advised Leonard Hartley about the development of his copper mine near Middleton Tyas and commented upon the pumping engine needed for the mine:

> 'By a Calculation I find an Engine whose Cylender is 42 Inches will work Three Sett of pumps of 12 Inch Diameter and 12 Fathom Deep and at a Moderate way of Working will Draw about Twelve Hundrd Hogsheads of Water Wine Measure in an Hour which is full Three Times as much as your Horse Engine can Draw when the Horses give a pritty pare. If you think that will do for you could wish to have a Regulating Beam at least 30'' Deep and 24'' Thick and 30 foot Long if possible to be gott that Length. My Reason for having it so long is I Design to make my stroke 2 foot longer than usual. Two Cylinder Beams at 24'' Square and 23 foot Long Each as soon as possible. Could wish to hear if you can gett Such 3 Trees and in the mean time will be enquiring about a Cylender and Working pieces but before I give orders for them Desires to know

---

[43] NEIMME: Brown/I/172 Leonard Hartley – William Brown 11th Sept. 1754; p. 175 WB-LH 25th September 1754; p. 175 LH – WB 4th Oct. 1754; p.163 LH-WB 25th June 1754; p. 175 WB-LH 25th September 1754; p. 196 WB-LH 22nd March 1755

if you think the Cylender 42 Inches be large Enough. A Boyler to Serve Such a Cylender will require about 20 Bolls of Coales a Day…your Daily expense of Coales will be 18s 9d if it is obliged to work the Engine Constant…There will be several pieces of Oak Timber wanting besides the Beam but as they are not very bulky I apprehend they may be got with no great Difficulty'.

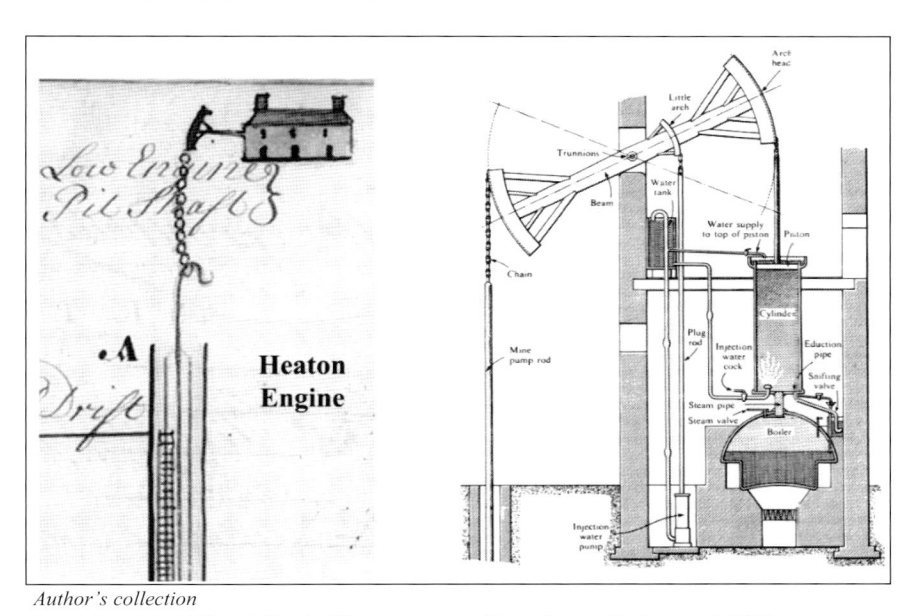

Author's collection

Fig. 10: A Newcomen Engine of about 1730

Acquiring the large timbers needed to construct an engine was a major problem and Hartley had the connections to help with this task. Earlier in 1753, he had written to advise his friend that some oak was available for building engines on Tyneside:

'I See last week the man who has the Oak beams all which is undisposed of and I find by him the Demand for….Timber is very Easy and he would gladly Deal for 3 or 4 on Moderate Terms. I think he would Deliver at Newcastle for 3s a foot or less as I have seen them and am pretty sure there can be no such had for a fire Engine in the north……Mr Fenwick's (one of Brown's enginewrights) coming here he might see them in Two Days time and fully satisfy you of what I say..To have them slabed to the Exact Squares you want where they now lye so that you will have no waste To pay for and there is a Number of them that will have wood to Spare in that respect….as to Elme for Pumps I have an offer of plenty as good as can be seen in any Country that you may be supplied with'.

There was a discussion about the merits of iron pumps as opposed to wooden ones and Brown commented:

'I find such pumps will cost you about £5 10s p. Fath. And wood pumps with hoops etc will cost you about £2 2s p. Fath. So considering the one will always Sell near the price cost and the other will Rott and Decay in a few years so in the Long main thinks the Iron as Cheap as Wood so Desires to know by Return of post which you will fix on'.

Acquiring these large timbers needed for the regulating beam and the supports for the cylinder was only part of the problem of building a pumping engine; getting a cylinder, the pumps and the boiler were the other major issues. Cylinders were available on Tyneside and Brown advised Hartley that:

> 'I have been seeing about a Cylender at Newcastle but finds none fitt for you but one and at present cannot find whether it is sold or no but will learn in a few Days and if it is not to be sold will order one with proper working pieces'.

Boilers were made from cast iron plates which were manufactured by firms such as Crowleys of Swalwell and Hawks of Gateshead principally for the production of salt pans – the large vats used to evaporate salt from seawater. Because of the availability of cheap small coals, salt making was a major industry along the coast from Blyth to South Shields in the eighteenth century. The plates were shaped and riveted together to make a boiler by the local blacksmiths. Carlisle Spedding, when seeking Brown's assistance for an engine at Workington Colliery, inquired 'if there be plenty of Good Boyler Salt pann plates and what they shall sell for'. [44]

Writing to John Robinson of the Friary Richmond regarding the erection of a fire engine at Grassington lead mine, Brown raised the problem of securing the timber and suggested that 'Least you may Mistake my Dementions or choose Such Timber as is not fitt…I think it will be prudent to send over my Engine Wright to the place who can…chose the Suitablest Timber of the Oake kind. When he has done that can give you Dementions of the Iron work to an Exactness and can have it all made to perfection at this place'. However, transport was a problem since 'The Roads So Excessive Rotten that it will be difficult to get Timber and other Materials Led …. We think it will be best to differ the Erecting of an Engine till Aprill or May when one may hope the Weather and Roads will be much better'.

Like many viewers, Brown became a part owner of the collieries in which he worked. Besides Throckley Colliery, in which he owned a half share, Brown had shares in both Shiremoor and its successor Willington, which were very successful enterprises. Willington Colliery opened in 1775, amid great rejoicing when the owners supplied a roasted ox, a large quantity of ale and a waggon load of punch. This winning made William Brown a major coalowner. His obituary, printed in the Newcastle Courant in February 1782, noted that 'yesterday, at his house at Willington, Mr William Brown, a considerable Coal Owner and principal viewer of the Collieries in this country; a gentleman greatly respected by his employers for his skill and integrity' had died. It was these personal qualities which explain William Brown's success and how he was able to rise from being a modest viewer at Throckley, to become an important coalowner and an authority on the science of mining, with a national reputation for designing waggonways and building steam engines. Although by the time of his death he was living at Willington House, William Brown was buried in the family grave in Heddon churchyard next to the children who had predeceased him. The family tombstone has an interesting coat of arms depicting three lions passant in bend between two double cotises (fig. 1). These were the arms of Browne of Beechworth, a family from South East England, which had become extinct in 1690. It would appear that this was a case of the nouveau riche, Brown of Throckley, appropriating an ancient lineage

---

[44] NEIMME: Brown/I/190 William Brown – Leonard Hartley 21st Jan. 1755; p. 152 WB-LH 29th June 1753; p.154 WB-LH 5th Aug. 1753; p.148 LH-WB 28th May 1753; p.8 CS-WB 4th May 1750; p.138 WB-JR 14th Oct. 1755; p. 140 JR-WB 13th Jan. 1756.

to enhance his family's status in society. The motto 'Suivez Raison', follow reason, is very appropriate for a viewer.[45]

William Brown was succeeded by his eldest son, also called William, who inherited his father's business interests and continued the development of the Tyne Basin by opening up Bigges Main, the neighbouring colliery to Willington, with his father's business partner, Matthew Bell, and his father's apprentice, George Johnson. William Brown Junior (1740-1812), through his mother's title, acquired the estates of her father, William Dixon, and assumed the name Dixon Dixon-Brown. He lived at Longbenton near the site of today's Ministry of Pensions. His son became Governor of the Hostmen's Company in 1826 and High Sheriff of Northumberland in 1827. Within a century, the Brown family had risen from being operators of a small landsale colliery in Throckley to become major shareholders in one of the largest coalmining enterprises in the world which earned them a prominent place in Tyneside society. This was in no small measure due to their personal integrity and professional skill as engineers.

However humble his early life may have been, in later life William Brown was a man of stature, 'the father of the trade' in Edington's words. A story in the Bell Collection at the Mining Institute supports this assessment and shows that William Brown was also a man with a very earthy sense of humour. The Bell papers record that Brown 'chewed tobacco and inside the lid of his box he had 'Cxxt and the Coal Trade' engraved'. One evening after dinner at Alnwick Castle, he 'took a quid of his favourite and was asked for one by his Neighbour who, on taking it, observed the Motto, and showing it to another, caused a laugh, which being remarked by her Grace, she insisted on seeing the Box, which request being backed by his Grace, it was handed up to him and immediately placed in her Grace's hands. She after examining it said 'I'll give you a Toast' … and she gave 'Willy Brown's Tobacco Box' which being drunk she left the Room – amidst thunders of applause! And 'Willy Brown's Tobacco Box' became the Standing Toast for many a day amongst the Coal Trade of the River Tyne.[46]

History has not been kind to William Brown and his treatment is in marked contrast to the celebratory status given to his neighbour George Stephenson. Yet these men had much in common: both were acknowledged by their contemporaries as masterbuilders of railways and steam engines, Brown building wooden railways and stationary steam engines in the eighteenth century while Stephenson built iron railways and travelling engines in the nineteenth century. Brown figures prominently in Dunn's 'History of the Viewers' (1811) and is mentioned in Robert Edington's 'Treatise on the Coal Trade' (1813). Edington made the interesting comment that Brown 'so far succeeded in the science of coalmining as enabled him to give lessons to others and made himself highly respected as the father of the trade'. William Brown was an engineer closely associated with some of the great names of the industrial revolution – men such as Abraham Darby, James Brindley and John Smeaton. However, there are no books to compare with the eulogy to Stephenson written by Samuel Smiles in his 'History of the Engineers' (1862) and the many other books that followed. Despite his undoubted achievements, William Brown has been largely airbrushed out of the annals of history. Hopefully, this study will do something to restore the esteem with which William Brown was held by his contemporaries.

---

[45] I am indebted to Andrew Curtis of the Heddon Local History Society for the photograph and information relating to the tombstone.
[46] NEIMME: Bell 14/304.

Chapter Three: The Coal Trade – William Brown's Business

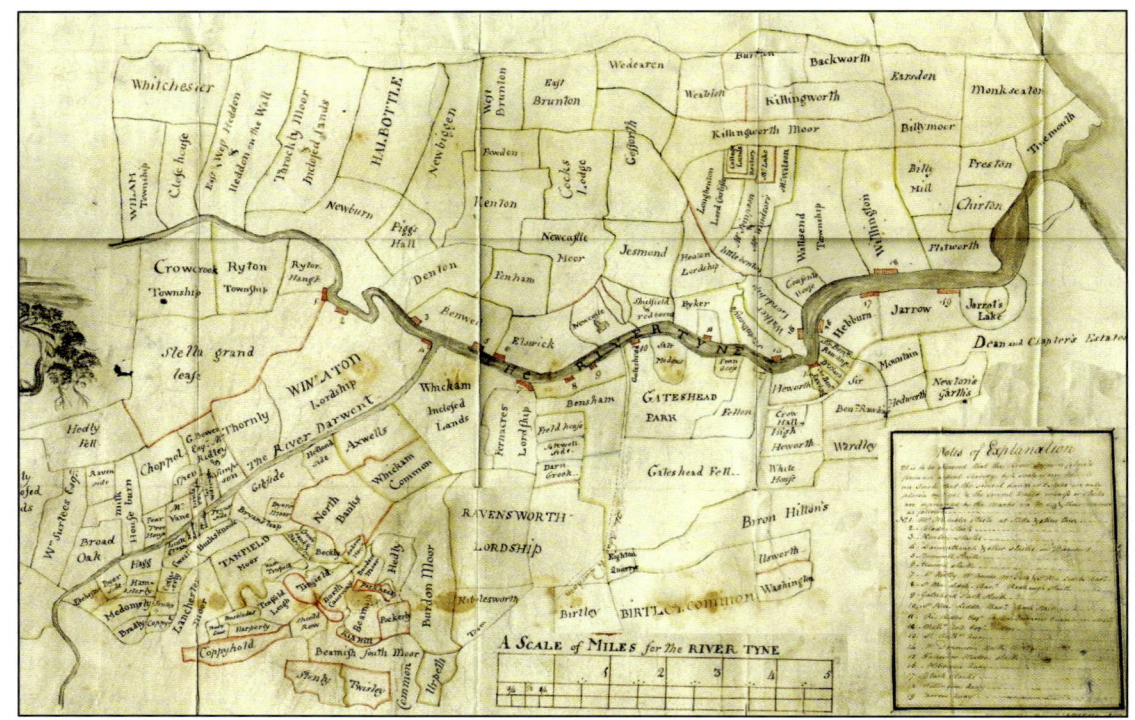

NEIMME:SR410.164WIL

Fig. 11: Wilson's Plan of the River Tyne 1754

William Brown's letters provide a very interesting and full account of the nature of the coal trade from the River Tyne in the mid eighteenth century; and, because they are written by a man at the centre of this business, they have a special value. There were three aspects of the coal trade: the landsale trade, the export trade and the seasale trade. The landsale trade sold to a local market and usually dealt in inferior coals which were used for lime burning, brewing, baking, rendering animal carcases and in the salt, glass, pottery and brick making industries. There are few statistics for this trade but it was substantial enough for the Montagus to employ a clerk of landsale at East Denton Colliery and was a significant part of the production at Throckley Colliery. The export trade to foreign countries was relatively small in the mid eighteenth century: it amounted to about 100,000 Newcastle chaldrons in 1789, half of which was sent to Holland. The seasale trade, which marketed the best quality household coal, was by far the largest and most lucrative part of the business: in 1788, it was 458,000 chaldrons. William Brown, writing to a client near Edinburgh about the selling prices of different types of coals from Throckley in 1755, noted that 'the best coals for London Markett sell at 14s, over Sea Small Coales at 8s, pan coales at 6s and some 5s 6d'.[47]

In Brown's time, two thirds of the region's coal trade were shipped from the River Tyne. Wilson's plan of 1754 (fig.10) shows the staiths of the main owners and the

---

[47] NEIMME: View book of George Johnson 1774 – 75 records that the output of Throckley Colliery in 1774 was seasale round 15,527 chaldrons, seasale small 1,987 chaldrons and landsale 2,792 chaldrons. Charles Beaumont, 'Treatise of the Coal Trade', p.21 lists the trade in 1789 as Dutch United Provinces 50,000; France and Flanders 20,000; Denmark 10,000; Hamburg 10,000; Sweden and Portugal 5,000; Russia, Norway and others 5,000. A similar pattern of trade is to be found in Edington's Treatise (1813) p.101 but by 1813 the total volume of trade had increased by 25%. Watson 22/8/404. Brown/I/245, William Brown to John Burrell, 26th April 1755.

ballast quays of Newcastle Corporation.[48] Curiously, there is nothing at Lemington. In the mid eighteenth century, most of the principal coalmines were west of the Tyne Bridge and keel boats carrying a cargo of about 21 tons were used to ferry the coal to the sea going ships, the colliers, which were largely moored at North and South Shields. This situation would change when William Brown opened up the Tyne Basin in South East Northumberland: then the colliers could be loaded with most of their cargo directly from the staiths at Wallsend, Willington and Whitehill Point.

Throughout the eighteenth century, the principal coalowners attempted to regulate the seasale trade to restrict competition and maintain high prices. This was to avoid what they termed 'a fighting trade' during which owners undercut each other to gain a market for their coal. In a letter to Carlisle Spedding, William Brown explained how the major producers were able to illegally rig the market to the detriment of the keelmen.

> There is a sort of Regulation or agreement entered into by most of the powerful Gentlemen in the Coale Trade which they oblige themselves not to sell their coales under such prices as is therein mention'd; and also that they are not to exceed the vend of a certain quantity; and also that they take their turns as to the vend alternatively one after another to vend a certain quantity each turn till the year determines; and in the case one or more Gentlemen had vended the quantity assigned him considerable time before the year determines, he or they may… not…vend more that Season but must throw his or their dealers upon such of the gentlemen concerned as is Short and has not vended the quantity assigned him. This is what we term the Contract and has been these three years past…justly observed; not withstanding the Express Acts to the Contrary…by the Regulation there is much profit arises at the Vending Ten Thousand Chalders as there is Thirty Thousand Chalders when there is a fighting trade, for when that is the case, one owner undersells another so that some sell cheaper than they work and then the Ships Masters make a fine time ont. There has been two meetings of the Gentlemen in order to Settle the said regulation for three years more but has not yet agreed'.[49]

The main problem appears to have been Brown's neighbour, Humble, who, in 1749, was trying to win Shiremoor Colliery near Billy Mill. The following year, in April 1750, Brown reported to Spedding.

> 'The coal trade in our river is in a bad situation such a one as has not been seen in the last century…nothing fixed as to a new regulation. In consequence of that every owner began to undersell another tho not in price fair but in giving extraordinary measure of Coales viz 9 or 10 Chalders of Coales in stead of 8 and received no more for them than for 8….This sort of dealing begun to lie hard upon the Keelmen so that they could scarce keep the Keel above water and labouring under Several Grievances before, they on the 19th March last made entire stop so that not a man of them wd go nor suffer any else to go up or down

---

[48] NEIMME: SR 410-164 Wil. The note of explanation reads: 1. Mr Humble's staith at Stella and others there; 2. Bladon staith; 3. Kenton staiths; 4. Derwenthaugh and other staiths in Darwent; 5. Benwell staith: 6. Elswick staith; 7. Mr Wortley, Mr Bowes, Mr Pitts and Sir Henry Liddle Bart; 8. Sir Henry Liddle Bart. Redheugh staith; 9. Gateshead Park staith; 10. Sir Henry Liddle Bart. Rock staith; 11. Richard Ridley Esq. Byker, Jesmond, Owesburn staiths; 12. Math. Bell Esq., Fellon; 13. St. Anthonys Quay; 14. William Donnisons North Birtley; 15. Walker or Heaton staith; 16. Hebburn Quay; 17. Black staiths; 18. Willington Quay; 19. Jarrow Quay.

[49] NEIMME: Brown/I/1 – William Brown to Carlisle Spedding, 13th January 1749

the river with any boats to carry coales excepting the Glasshouses and Saltpans….The fitters sent 20 or 30 of the principal Keelmen to goal, being bound servants, in order to frighten the rest but to no purpose. It was thought that the Rabble part of the Keelmen wd endeavour to risque their going to jail and raise a mob or riot, so as a military force might have been turned upon them and obligd them to work, which the magistrates had called into Town for fear of a great disturbance, but instead…they…go with the utmost civility….Cash grows scarce with the coalowners, nothing is heard but complaints….The River looked dejected or rather deserted nothing is moving except a few wherrys going up and down instead of Keels to the No of Five or Six Hundred….the Harbour is full of ships but empty ones….Consequently, Sunderland doing a fine trade and would have done much more but the wind is against them'.[50]

On 12th July, he reported to Spedding that 'the masters then in Harbour sent their Sailors to work the Keels up the river and, being assisted by Pittmen and waggonmen down, got loaded so…the rest took the hint and the Keelmen went to work by degrees'.[51] Clearly, there was no solidarity amongst the working classes of Tyneside on this occasion. A price war ensued between the principal producers, Lord Ravensworth, Bowes and the Wortley – Montagu family on the one side, and Lord Windsor, Matthew Ridley and John Simpson on the other. Finally, an agreement was reached and the allocation agreed for each colliery is shown below.

Sept 1753

Estimate of the quantity of Coals which will be vended by the several Coal Owners in the River Tyne this year by chaldrons

Lord Windsor	Lanchester Moor, Crookbank, Collierley	45,000
Lord Shelburn	do do	11,000
Jno. Simpson	Pontop, Lanchester Moor, Lands	22,000
Wm. Pitt Esq	Tanfield Moor	20,000
Matt. Ridley Esq	Tanfield Moor Edge*, Byker*, Lands	35,000
Mr Geo. Silvertop	Brockwell Grand Lease	12,000
Mr Jno. Humble	**Newburn Moor*, Risemoor*, Bradley Moor**	**12,000**
Mr Jno. Beaumont	**West Denton***	**7,000**
Mr Wm. Barkas	**Throckley Moor**	**8,000**
Mr Matt. Bell	Brockwell	3,000
Sir Thos. Clavering	Byermoor	2,000
Mr Robinson	Gellsfield	2,000
Mr Liddle	Saltwellside	2,000
Mr Kyle	Gateshead Fell	1,000
Mr Hodgson	Risemoor* and Quarry House	3,000
Lord Ravensworths	Family Colly at Ravensworth*	25,000
G. Bowes Esq	Family Colly at Northbanks	7,000
Partnership	Benton*, Park*, Burdon, Hedley Moor, Park House	100,000
Including overseas coals suppose the vend to be		315,000
Francis Blake Delaval	Hartley*	10,000
Matt. Ridley Esq.	Plessey	15,000
Mr Greathead	Blackclose	400
	Total	340,400 [52]

Such documents reveal the relative importance of the different collieries and the different areas of the coalfield at a given point of time. In 1753, the dominance of

---

[50] NEIMME: Brown/I/4 William Brown to Carlisle Spedding, 30th April 1750.
[51] NEIMME: Brown/I/9 William Brown to Carlisle Spedding, 12th July 1750.
[52] NEIMME: ZA/11/279. (* 9 of 31 collieries known to have one or more fire engines).

Tanfield Moor, served by the Tanfield Way to Dunston, and the Derwent Valley, served by the Western Way to Derwenthaugh, is evident but this would change as a result of William Brown's conquest of the Tyne Basin. Throckley and Newburn Moor are the only collieries supplying the seasale trade from Lemington mentioned; but in the next quarter of a century, although Throckley and Newburn (Walbottle) remained the most important, they would be joined by Holiwell and Wylam as the later vends illustrate:[53]

	1778	1779	1780
Throckley	14,686	11,517	11,528
Newburn	13,186	10,871	26,121
Holiwell	10,432	7,335	4,989
Wylam	11,327	8,753	8,340

Unlike the previous document, which recorded the colliery's allocation, this table records the actual figures for the seasale trade; but this was not the entire production of the colliery. Landsale coals are not included and the common practice of giving the captain of the colliers over measure to promote sales is not reflected in these figures. Henry Masterman, one of the owners of Throckley Colliery, complained about 'those large allowances to Cpts of Ships' but William Brown explained that 'allowances certainly sink'd the profits of the Colly, yet there is at some times an absolute necessity to do so, for when our coales is Low at market and plenty of sorts to gott that sells higher, the masters will not load unless they have prices Insured, which I find is the case with most Collys in the River'.[54]

Wylam Leadings 1764-5

			Weekly
May 22nd	1,354	(3)	*452*
June 19th	1,641	(4)	*410*
July 24th	2,149	(5)	*436*
August 28th	1,780	(5)	*270*
September 11th	927	(2)	*463*
October 9th	1,560	(4)	*390*
November 6th	1,498	(4)	*374*
November 13th	363	(1)	*363*
December 31st	2,490	(6)	*415*
January 15th	379	(2)	*190*
February 12th	1,369	(4)	*342*
March 12th	1,633	(4)	*408*
April 9th	1,555	(4)	*388*
April 30th	1,268	(3)	*423*
Total		**19,966**	

Newbiggin (Holiwell) Leadings 1764-5

			Weekly
Dec 15th - Feb 6th	1,496	(5)	*300*
March 20th	1,911	(6)	*320*
April 24th	1,499	(5)	*300*
June 5th	1,484	(6)	*247*
July 10th	1,507	(5)	*301*
August 21st	1,965	(6)	*327*
September	Strike	(5)	
October 30th	1,043	(4)	*261*
November 27th	1,689	(4)	*422*
December 18th	1,846	(3)	*615*
(5) = 5 weeks			
Total for 1765	**14,440**		

The pattern of trade during the course of a year is shown in the figures of leadings which were collected by the viewers as the basis for calculating wayleave payments to the landowners. The figures for Wylam and Holiwell were recorded by Christopher Bedlington to assess the wayleave rent due to the Earl of Carlisle: the route of the Wylam Waggonway ran through his lands in Heddon on the Wall and the Holiwell Way

---

[53] NEIMME: Watson 2/11/2. The figure for Duke of Northumberland in 1780 (26,121) included Preston Moor.

[54] NEIMME: Brown/1/128 Henry Masterman to William Brown, 12th January 1753 and p. 129 William Brown to Henry Masterman, 10th February 1753.

ran through his estate at Newbiggin.[55] There was a regular flow of traffic along these waggonways throughout the year but there was a peak in the summer as the market prepared for the autumn sales and in December before winter set in and limited sailings. It is interesting to note the effect of the strike of 1765 when the miners refused to accept the alterations to the annual bond which attempted to tie the men to a particular colliery. The owners, faced with the difficulties of retaining the workforce at a time when the industry was expanding, were insisting that the men were not to be allowed to leave without a discharge certificate which the men (and interestingly also the Montagus at East Denton) regarded as tantamount to slavery. The men won their fight.

Industrial action amongst the miners and keelmen was not the only cause of stoppages in the coal trade. In 1755, Brown reported to Masterman that the Thistle Pit in Throckley had opened and the seam contained four foot of 'the finest and best coal I have ever seen yet at ye Colliery'. However, his next letter contained disturbing news that 'We have had no Trade in our River these 16 or 18 Days past most of the sailors being pressd and run out of the way so that our staith is full.' The bulk of England's mercantile fleet was employed in the coal trade down the east coast and the collier fleet was the prime recruiting ground for the Royal Navy especially in times of unrest. The outbreak of the Seven Years War in Europe in 1756 was such an occasion. William Brown informed Henry Masterman that 'The Coale Trade in our River is Intirely at a Stand and has been so since the 6th Inst. Scarce a watter Man of any kind Dare look out at Newcastle they Impress all the Watter Men they can meet with that has the least appearance of doing Serviss and it is corrintly reported that Next week Landmen will be Impressed…A Report prevails here that our Enemies The French will certainly Land at Newcastle…Has got a new Waggon Way Laid through the South part of Throckley Estate to the respective new pitts and now to my mortification can gett no Coales Vended the Steathe is quite full and a great quantity at the pits and from what I can see there is no likelihood of the trade being opened.' During the war with France, the collier ships travelled in convey with a naval escort down the east coast for it was a popular tactic of the nation's enemies to attempt to rob the capital of its fuel supply by attacking the colliers from the Tyne and Wear.[56]

In the eighteenth century, only members of the Hostsman's Guild were permitted to trade from the port of Newcastle and coalowners hired a fitter, or shipping agent, who was a member of that guild, to market their coal. Mr Barkas was the fitter for Throckley Colliery in which he also owned a quarter share. The fitters owned the keel boats, the lighters which shipped the coal to the colliers generally moored at Shields. They hired the keelmen and were responsible for obtaining all the documents relating to the cargo and paying the charges imposed by the port. As the middle men, they guaranteed the delivery of coal to the ship's captain from whom they received payment for the cargo; and they also guaranteed payment to the owner. The ship's captain transported the coal down the east coast, a journey which took between four and five weeks since sailing ships were heavily dependent upon the weather. About 40% of the coal was sold at

---

[55] Source NEIMME: East 10a p.26 and 28; East 10b p.28.

[56] NEIMME: Brown/I/308 William Brown to Henry Masterman, 25th February 1755; p. 316 William Brown to Masterman, 5th June 1755; 340 William Brown to Henry Masterman, 16th March 1756.The waggonway referred to is probably the line from Heddon running in a north, north easterly direction past Throckley House and the pits Dyke, Bounder, Ash and Knab.

Kings Lynn in East Anglia and ports along the south coast, but London accounted for the lion's share. At London, the off-loading and marketing was controlled by coal factors, about fourteen in total, who met at the Coal Exchange in Billingsgate three days a week – Monday, Wednesday and Friday. They arranged for lighters to be available to ferry the cargo ashore, they organised the paperwork for the national and local taxes levied on the cargo, and negotiated the sale of the whole cargo to a principal buyer, who sold the coal on in smaller quantities to lesser buyers. These coal factors were the agents of the Tyneside coalowners.

The man responsible for marketing Throckley coal in London was Captain John Biggins who had an office in the Salutation Inn at Billingsgate near the Coal Exchange. Captain Biggins regularly sent William Brown reports on the reception of his coal by the Exchange. Coal, like fine wines, was marketed by the estate and it was important to establish and protect the reputation of the estate. This is shown in an interesting letter from Henry Masterman concerning the ill effects of a contract which the colliery had made to dispose of the inferior small coals in London.

> 'Capt Biggins was with me this morning to inform me yt my Namesake had made a Contract with ye Chelsea Waterworks to supply them with 500 Chalders of our small Coal from Throckley Colliery; but he advises by all means not to let any of these small coal be sent to this Town upon any Acct whatever; for he says we shall sure Risque of having a Discredit brought upon the whole Colliery by it; for he apprehends…it be believed that all our Coal is, or however may be, mixt with this small stuff; and therefore would have the small sent elsewhere.'[57]

For his part, Brown feeds Biggins with information about the situation at Throckley and the other collieries supplying the market from Tyneside. This information would enable Biggins to understand Throckley Colliery's position in relation to its competitors.

> 'The new pit … did not answer expectations… several Troubles and Hitches rendered the Coales very mean….another new pitt … will be at work in Six Weeks. That pitt will far exceed anything you have had from us in Round Coale and quality. There is very little to do in our port at present several having Stocks of Coales at Steaths. At present Longbenton is working very hard and does not mix their coale so much … they work Dayly about 240 London Chalders – Byker works about 110 London Chalders Dayly – Park has been much plagued with additional feeders of water when they can work …they work Dayle 120 London Chalders. Longbenton and Byker has few coales ready wrought but Park has a great many – West Denton is quite full at the Steath and Some at pitts – The Leaser and the Leasees of that Colly. has got a Chancery Suit among hands so is great confusion – Mr Humble's Newburn Moor makes a poor figure. They work very few and these not merchantable: they are soft and small and full of bands of splint. – Rise Moor has done little since Christmas on account of a dispute twixt Mr Humble and his partner Mrs Hodgson. Humble's Bradley Moor works pritty well but does not send one Chaldrons out of three to London Market they work so Small and Tender. Grand Moor has great quantitys both at pitts and steaths. They work a pritty quantity but leaves a great many Small at

---

[57] NEIMME: Brown/I/278 Henry Masterman to William Brown, 24th May 1753.

pitts, their quality is Tender and full small. – Pontop's has got most of their foreign Heaps Led to Steath so that in a little time they will only have coales as fast as they can work them. Tanfield Moor is vending very fast and it's expected they will get their quantity before Michaelmass. I think they come smaller this year than they have hitherto done. Tanfield steath is pritty full notwithstanding they are 8d per Chaldron lower than other Collys.'[58]

Captain Biggins replies are often blunt and uncompromising as befits the marketing department.

'As to Throckley Moor … the first two ships were reported to me Round both by your people and the Master but proved to be Scarse Housekeepers Coales; the two last… I apprehend are worse…now we are in as much Disgrace as ever, so must beg the next ships you send may be of the best coal you can get that I may endeavour to raise our Character before you be able to send a large quantity. I also must take the Liberty to Recommend your peoples sincerity with the Masters which is the greatest complaint alleging that they are promis'd to be loaded in 6 or 4 days when are kept five times that time at every opportunity.'[59]

In his next letter Brown was anxious to reassure Captain Biggins.

'This Day I expect the Isabella, Cpt Jno Stevenson, will sail from hence for your place with a very good bulk of our Throckley Moor Coales. He was at the Colly and can partly acquaint you with our situation. We have got the new pitt Down…and dayly increases our quantity wrought out of her. At first She wrought very wet but now is turn'd very Dry and the Coales exceed both in quality and roundness and in colour any we ever wrought out of Throckley Colly. When any small happens to be wrought (which must happen in all Collys) we now sell them as Small for oversea and they begin to be much lik'd so we have got a good demand for them. I hope in the future that you will not have room for to complain for I am convinc'd you will have one bulk of Coales come like another and may depend on there being Round and well Sorted.'[60]

The letters refer to Throckley Colliery but they illustrate the problems faced by coalowners throughout the region: the difficulties of agreeing a regulation between themselves, usually referred to as 'the vend', to strengthen their position on the London market; of recruiting labour, dealing with industrial disputes, maintaining production and protecting the colliery's reputation; of keeping the essential transport personnel, the waggonmen, the keelmen and the ships' captains, content. The actions of the press gangs and the dangers arising from war in Europe added further worry. Clearly, running a colliery in the eighteenth century was not for the faint hearted!

---

[58] NEIMME: Brown/1/222 William Brown to Cpt John Biggins, 2nd June 1753. A London chaldron was 28 cwt a little more than half the Newcastle chaldron of 53 cwt.

[59] NEIMME: Brown/I/278 Cpt John Biggins to William Brown, 13th June 1753

[60] NEIMME: Brown/I/221 William Brown to Cpt John Biggins, 22nd September 1753.

Chapter Four: The Work of a Colliery Viewer – Brown's Profession

The mystery of winning and working collieries was a skill owned by certain families in the northern coalfield; and the description which Amos Barnes gives of his training is in all probability typical of many other families of viewers. The fact that the account was given to establish his credibility in a court case adds weight to the evidence. His father, Jonathan Barnes Senior, was the agent and viewer for Sir Henry Liddell, a founder member of the Grand Allies. Amos 'was introduced by his Father…from his Childhood in the Knowledge of Collierys until he was 21 years of Age when he was appointed Agent and Viewer at Heaton and Long Benton Collierys to the present Lord Ravensworth and Partners'. These were two of the most technically advanced mines in the world in their day and both were equipped with the revolutionary Newcomen pumping engines. Amos Barnes 'continued for the space of 23 years…and during all that time he had the Chief Management and direction of the winning and working the collierys….At the expiration of 23 years quitted…and was employed as Agent and Viewer for Matthew Ridley Esq. for his Colliery at Byker and continued in his Service for 3 years'.

However, his experience was not confined to his principal place of employment for 'hath been employed as Viewer for several other Collierys in Northumberland and Durham ever since being first appointed Viewer at Heaton and Long Benton and has by that means acquired a Knowledge and Skill of Winning and Working Collierys'.[61] Amos Barnes' son, Jonathan Barnes Junior, would follow in his father's footsteps; as did William Brown's sons, William and Richard; and William Bedlington's sons, William and Christopher. The pattern of an apprenticeship, very often under one's father, followed by an appointment at a particular colliery, then consultancy work at numerous other collieries in the region and elsewhere, appears to have been a common career ladder. Edington claimed that 'William Brown…was bread from his youth in colliery affairs'. His father was probably his mentor but whether his neighbour Richard Peck, his relative William Newton, the Grand Allies men Jonathan Barnes Senior, Amos Barnes and Nicholas Walton, or some other viewer was his master is not known.[62]

Throughout the eighteenth century, this fraternity of viewers worked in a region where there was an enthusiasm for mathematical knowledge and an interest in scientific enquiry. Although contemporary reports emphasise that viewers were trained on the job, there was no shortage of very able schoolmasters in the area, neither was there a lack of text books, to provide the theoretical background in mathematics and science to underpin their practical experience. Most of their employers, the major coalowners such as the Liddells, Ridleys, Lambtons and Montagus, also had residences in London, where they became aware of the work of The Royal Society and in particular the ideas of Isaac Newton. Having a dual residence facilitated an exchange of information between the capital and the North East: for example, Edward Montagu of Portman Square, London, and East Denton Hall, Newcastle, was a distinguished mathematician and a fellow of The Royal Society; and this exchange was not a one way exercise for Montagu contributed articles on mathematical subjects to the influential 'Gentleman's Magazine' and others from the region wrote on technical matters such as the design and operation

---

[61] NEIMME: Forster 1/4/43. Transactions Volume 81 (1930-31) T.V. Simpson Old Mining Records.

[62] Edington R., A Treatise on the Coal Trade, London 1813, p.123.

of the coal waggon.[63] Edward Montagu befriended William Emerson of Hurworth near Darlington, who had been a pupil at Trinity House School in Newcastle, a distinguished centre of mathematical learning. Emerson became a prolific writer of mathematical text books which earned him an international reputation. Nor was Emerson's achievement unique: a contemporary, Charles Hutton of Heaton, had a more dramatic rise to fame. Born the son of the overman at Longbenton Colliery, he failed as a pitman because of an injury to his arm sustained in childhood and became a schoolmaster. He ran a very successful school at Stote Hall in Jesmond, Newcastle, which was noted for its teaching of mathematics, before being appointed professor at the Royal Military Academy in London, one of the most prestigious posts in the country.[64] Charles Hutton would acknowledge his debt to Tyneside through his support for the Literary and Philosophical Society which was a meeting point for colliery engineers such as William Thomas, John Buddle and William Chapman. Such was the cultural climate of Tyneside in which aspiring viewers operated; and many, like Edward Montagu's man, William Brown, would make their own contribution to this scientific knowledge base.[65]

In the early 1760's, Edward Montagu received a report, probably from William Newton, which advised him upon the workforce needed to operate his colliery at East Denton. This document contains a rare description of the role of the viewer:

> 'An Agent is necessary that is well acquainted with Collierys as a viewer who is to View the Colliery & look to the Engine & Staith also & keep the Accounts with a Clark for his Assistance. The whole cannot be less for the Viewer than £60 & £20 a year more for a Clark or other Assistants.'[66]

The viewer was the man with the technical expertise to win the colliery and afterwards to wage a constant battle against the emerging geology, when faults in the earth's crust and water from the strata caused havoc with the progress of the mine, and threatened its very existence. He was responsible for all the machinery: the horse gins, wind gins, water mills and the fire engines for pumping and winding, the waggonways for transport and the staith for storage and shipment. But the viewer was more than an engineer and William Brown draws this distinction in a letter to Carlisle Spedding referring to Michael Menzies as 'our Scotch Engineer who is now very full of a new contrivance (viz) to apply the power of the atmosphere to Raise water from any Deepness without the help of Beams, Chains and Spears'. Spedding had previously referred to Menzies as 'the Scotch gentleman who got the patent for winding Coales without Horses' which he had installed at Chartershaugh Colliery, on the north bank of the River Wear, in 1753. His experiments were to culminate in the building of the double water wheel, driven by the water from the pumping engine, to raise coal. It was claimed that this haulage engine could greatly reduce to cost of horses at a colliery. Furthermore, Menzies was the first to introduce inclined planes to improve transport underground. He was clearly a talented engineer but Menzies was not considered to be a viewer.[67]

---

[63] See letter to the Gentleman's Magazine by T.S. Polyhestor of Chester-le-Street of 21st December 1763.
[64] Horsley, P.M., 'Eighteenth Century Newcastle', p.25-52.

[65] In 1792, when advocating the creation of a literary and philosophical society, the Rev. Turner argued that one reason for Newcastle being a suitable place was that 'their existed persons who are employed as viewers and are capable of supplying better information than can be obtained any other way' about the two great products of the area – coal and lead. Watson R.S. 'History of the Lit and Phil' p.36.

[66] Northumberland Record Office: Sant/Gen/Est/1/4/2/10. The full document is discussed later

[67] NEIMME: WBLB I p.31, 261, 264; Transactions XV p.207; Taylor, Archaeology of the Coal Trade, p. 194.

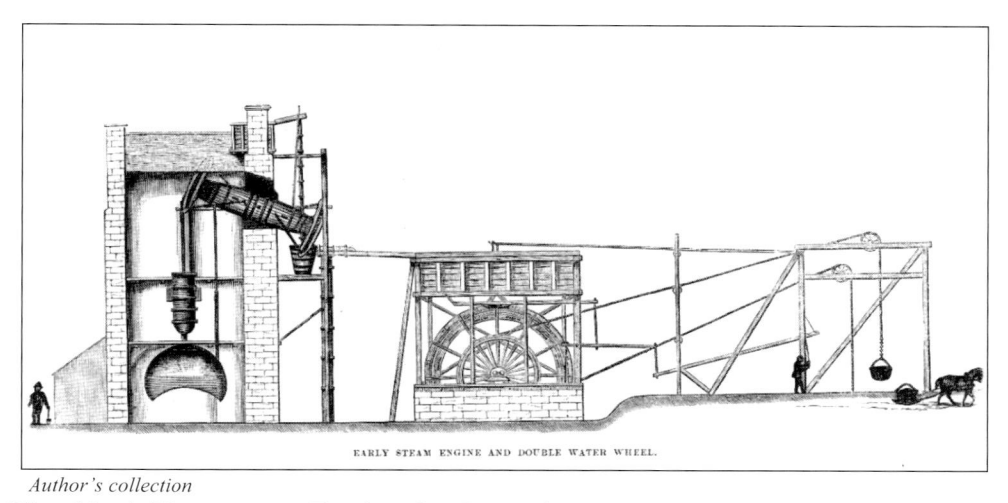

Author's collection
Fig. 12: A Newcomen Engine for Pumping and Water Wheel for Haulage

In addition to having engineering ability, the viewer was required to be a competent surveyor, who could 'line the pit', that is make plans of the workings underground. He was expected to provide financial advice to the owners on the viability of the business and to recruit, manage and retain the workforce needed to operate the enterprise. The viewer was also responsible for the farms which supplied feed for the horses used by the colliery both above and below ground. At Denton, the viewer also had oversight of the ancillary industries – what Elizabeth Montagu called 'the bricks, tiles, tar manufactory etc. going on at the waterside'. For these reasons, the resident viewer was the highest paid member of the workforce, earning almost twice as much as the next best paid employee, the enginewright, who was another essential member of staff. The viewer worked closely with the men underground: he was responsible for their safety and was usually the first man on the scene if an accident occurred. Viewers were often held with great respect by the men and sometimes even affection. Mrs Montagu wrote to her friend Elizabeth Carter that 'the Pittmen tell my servants that Wright (her viewer) is very humane and kind to them'; and the obituary of his successor, Ralph Allison, published in the Newcastle Journal in July 1770, expresses the same sentiments.[68]

> 'Friday, died in the house of William Archdeacon Esq., in Newcastle, aged 39, Mr Ralph Allison, viewer to the Hon. Edward Montagu, Esq , in which station he acted with the most indefatigable care and strictest integrity…his humane behaviour to the men under him gained their esteem and love.'

However, William Newton is describing only the most basic type of viewer, the resident viewer, and there were other viewers, such as William Brown, who were involved in a wider range of work, both within and beyond the region. The resident viewer usually had a more senior man, who had oversight of a group of collieries, to supervise his work. At Elizabeth Montagu's colliery in East Denton, for example, William Brown supervised the work of Jonathan Wright in the 1760's; and Jonathan Smith monitored the work of William Thomas after 1787. The royalty owner also employed a viewer, generally a man of wide experience, to check that his tenant was working the colliery properly, with the long term interests of the owner in mind, and not recklessly, for a quick profit. Nicholas Walton and John Boag were the check viewers for Greenwich

---

[68] Huntington Library: MO 3173, Elizabeth Montagu to Elizabeth Carter, 16th January 1766.

Hospital and monitored William Brown's work at Throckley Colliery. When strategic decisions had to be made it was not uncommon to commission a report from a group of four or more viewers. Mrs Montagu commented to Lord Bath that 'we must consult all the Sages of the World below concerning the colliery' which referred to Edward's meetings with William Brown, Nicholas Walton, William Newton and other viewers prior to the winning of East Denton Colliery.[69] It is not surprising that coalowners sought the opinions of several leading viewers before acting for a large amount of capital was needed to win a mine before any returns were made from the sale of coal. As Elizabeth Montagu commented to Lord Bath on winning East Denton Colliery:

> 'I believe we have open'd a noble source of future plenty but it is present poverty. We are at present the poorer a great deal, for a mine at first opening has a prodigious swallow; when it begins to disgorge it makes noble amends.'[70]

A man's personal qualities were at least as important as his technical skills as is illustrated in William Brown's letter recommending Edward Smith of Houghton to a coalowner: 'I know nobody so proper to serve him as Viewer…he has both judgement, resolution and honesty'.[71] A great deal rested upon a man's reputation for integrity and discretion as Elizabeth Montagu vividly described in her account of a meeting at East Denton Hall in another letter to Lord Bath in 1763.

> 'I ventured to go down stairs yesterday for an hour to hear the report and view the calculations of the colliery and the plan for working it by the man of the greatest reputation and skill in this Country. This report they cannot make to any one but the owners for it is held as the most sacred thing imaginable and a Viewer who was to tell what he observed in the lower regions of any other would be ruined.'[72]

William Brown was well established as a viewer when, in 1758, Edward and Elizabeth Montagu travelled north to take possession of their inheritance – the estate of John Rogers, the last member of this major family in the Tyneside coal trade. The Montagus stayed at Carville Hall, Wallsend, because East Denton Hall was infested with rats. William Newton had been John Roger's viewer but after his death in 1762, Edward Montagu appointed William Brown, whom he held in very high esteem: 'In the science of collierys …he is superior to any man', Edward wrote to his wife in January 1764.[73]

Senior viewers were also asked to arbitrate in disputes between owners. Frustrated by the lack of progress at the Court of Chancery, in November 1774, the parties involved in a dispute relating to the future working of Birtley Colliery decided to speed matters up by agreeing to abide by the decision of three arbitors 'Christopher Fawcett Esq., William Wilson Esq. and William Brown Gentleman'. The arbitors reached the decision that William Peareth was not working South Birtley Colliery in the correct manner and outlined a programme for the proper management of the enterprise. The

---

[69] Huntington Library: MO 4592, Elizabeth Montagu to Lord Bath, 23rd October 1763.

[70] Huntington Library: MO 1438, Elizabeth Montagu to Lord Lyttelton, 4th July 1765

[71] NEIMME: Brown/II/25, William Brown to C. Swainton 27th February 1765.

[72] Huntington Library: MO 4601, Elizabeth Montagu to Lord Bath, 20th November 1763.

[73] Huntington Library: MO 2476, Edward Montagu to Elizabeth Montagu, 5th January 1764.

viewers had resolved the owners' dispute and no doubt saved them a considerable sum in legal fees as a consequence. [74]

Like Amos Barnes, Brown's sons, William and Richard, served an apprenticeship under their father's guidance as did the Bedlingtons and Gibsons. Afterwards some of his apprentices became part of his consultancy and the articles of agreement between William Brown and 'Christopher Bedlington of Shields Row in the parish of Tanfield', dated 8th December 1762, do survive. They illustrate what was expected of a member of Brown's staff. During the three years of the indenture, he was required to 'dwell with Continue and Serve the said William Brown…as his Servant and diligently and Faithfully according to the best and utmost of his power Skill and Knowledge exercise and employ himself and do and perform all such Service and Business whatsoever as well relating to the business of a Viewer of Collierys which the said William Brown now hath'. But that was not all for he was also required to attend to 'other Business and Matters and things whatsoever as the said William Brown…shall from time to time Order, direct and Appoint to and for the most profit and advantage of the said William Brown'. In all these matters he 'will keep the secrets of the said William Brown and be just, true and faithful to him in all Matters and things and no Ways wrongfully detain, imbezil or perloin any Moneys, Goods or things whatsoever of or belonging to the said William Brown'. Christopher Bedlington was required to 'keep just and true accounts…of all Moneys received and paid and of all other things whatsoever relating to the Business which shall Come or be Committed to his Care, Management or disposal and from time to time pay all such Sums of Money into William Brown's hands'. He was required 'to give true and fair Accounts of all his Actings and doings in the said Employment…without Fraud or Delay'. Bedlington's salary was 'Twenty Pounds of lawful Money of Great Britain by equal half yearly payments on the Twelve Day of May and the Twenty Second Day of November'. In addition, he was to be provided with 'sufficient meat, drink, Washing and lodging'; a Horse to ride upon for the more Expeditious Transacting of Business…together with an Allowance of such reasonable Expenses as shall necessarily expend for Refreshment or otherwise'. [75]

Fortunately, some of the view books of three of his assistants survive. The records of Christopher Bedlington, William Gibson and George Johnson provide a glimpse of the extent of Brown's influence, not only in Northumberland and Durham, but also in Yorkshire, Nottinghamshire and further afield in Scotland and Ireland. They also illustrate the sort of work undertaken by a colliery viewer in the eighteenth century. [76] These three apprentices were workaholics but whether through choice or compulsion is not known. For example, Gibson's diaries, which cover over seven years, reveal that he was at work everyday except one and never had a holiday, not even at Christmas. In 1771, he writes, 'being Xmasday Was in the office allday'. On Christmas Day 1775, he was much more active. 'On ye morng: I walkd to ye Engine wch was going and ye water near down and nothing else doing at any of ye works...I returned to Breakfast and afterwards rode wh Mr Richd. By way of ye moor etc to Benton and took ye leadings for this year….at noon I walkd with Mr Richd to Willington (shot 2 Ratts) had a bottle of strong Beer in his new house (by way of Hansel). We spent most of the evening at Cards'. Mr. Richd was William Brown's youngest son. Apart from the odd visit to the

---

[74] NEIMME: Forster 1/4/324
[75] NEIMME: Watson 3/106/1-2
[76] NEIMME: Gibson Diaries; East 10a and 10b Bedlington's Diaries; Johnson Diaries in the private collection of Steve Grudgings.

theatre, Gibson's principal recreational activity appears to have been horseracing. On Saturday 29th June 1771, he 'was in the office till past 5 o'clock and then went to the races where saw the Ladies purse of £50 won by Mr Hutton's Black colt Dormouse'; and on Monday 29th July 1771, he 'called at Lemington Hopping where saw 3 heets run by ye waggonhorses (for a saddle) and was won by Forster's horse of Newbournhall'. William Gibson was also fond of a drink: on Sunday 28th July 1771 he recorded that 'it being the hoppings drank pretty freely'; while on Wednesday 23rd 1774, he left Byker Colliery at four and 'came home by way of Heaton Hall and got too much drink'. It was a windy and showery night when he got home at eight o'clock.

Not all of their work was in the North East. Bedlington and Gibson accompanied William Brown on visits outside the area. On Friday September 13th 1771, Gibson set out for Fife travelling across the Firth of Forth by boat from Dunbar. At Dunfermline, he bought Brown's daughter, for whom he appears to have had an attraction, ten yards of ribbon and, being a diplomat, he also bought some for her mother. Gibson did not return to Newcastle until December 17th. Between August 29th and October 14th 1772, he was in South Yorkshire at Frystonhall Colliery near Ferrybridge. For his part, Christopher Bedlington records work at Aston Colliery belonging to the Earl of Holderness in February 1766; and in the following month he was at Mr Jones' colliery in Ireland. These men were engineers in the most technically advanced coalfield in the world and it is therefore no surprise that they should be called upon to take their skills outside the region. Later, engineers from the Great Northern Coalfield would travel to Europe, America, Africa and elsewhere in the world.

The viewer was responsible for the management of the workforce and industrial relations feature prominently in their writings. The miners signed an annual bond committing themselves to a particular colliery for twelve months. This was often a time of tension as both parties manoeuvred to secure the best deal. Bedlington visited Heworth Colliery on October 14th 1767 during the binding period. He found 'the pitmen are all idle and will not be bound… told Humble and Goodkid not to give any more than one shilling each householder and 10s 6d each to every non householder' as binding money. To break the bond was a criminal offence and employers went to extraordinary lengths to enforce this contract and retain their workforce because, in a highly capitalised and expanding industry, at a time when skilled workers were in short supply, the future of the enterprise depended upon maintaining production at an adequate level. Gibson spent a week in midwinter hunting down pitmen who had absconded from Walbottle Colliery in 1771.

Gibson recorded trouble with the seamen in March 1775 noting on 16th that 'nothing doing on the River ye Sailors yet steaking'; and, more alarmingly, on 20th that 'this evening ab 200 Sailors came in a riotous manner and allarmd ye house, but did no further hurt – wh obligd us to have some men by way of guarding the house at night…and day'. This was probably his mother's house at Tynemouth.[77] In April 1780, between Monday 10th and Thursday 20th, William Gibson had to deal with a strike of miners at Washington Colliery caused by the introduction of new technology – rollies to replace sledges underground. His diary provides a lengthy account and gives an insight into the viewer's role in managing labour relations. On the 10th he proposed a compromise: that 'two indifferent men should be chosen to see the difference between

---

[77] His father Thomas had a foundry in Tynemouth and his brother John was the surveyor of the well known map of the coalfield first published in 1787 and reprinted with amendments in 1788.

ye sledges and rollies and their report with my own trials should determine it'. However, this did not satisfy the men and the strike continued. On the 20th Gibson resolved upon action.

> 'As the pitmen still continued streaking, I was determined to have something done. Therefore ordered ye Constable to execute ye Warrant got yesterday which he did about 6 on morning upon which the place was immediately in an uproar and they all assembled (men, women and children about 120 in number) in a very riotous manner. They then came and I had a meeting with them. I plainly told them that I would not be any longer put of, and insisted on them either promising to go quietly to work (as proposed on the 10th) or to go before Mr Ilderton (the magistrate). But they refused; saying no Constable should take a man of them. I then saw there was no dealing with them in a reasonable way. Therefore told them I would be under necessity of taking such steps as would not be to their advantage. I then took with me ye Constable and went to Westoe, saw Mr Ilderton and laid the whole matter before him. He saw it very clearly and gave it as his opinion the men were quite wrong and ordered another assistant Constable from Shields (named Skipsey). If the men would not then go before him, he said he would send a Military force… I returned and got home about 3 after noon. Had a meeting with ye men and found they had altered their foolish resolutions and said they were willing to work without any consideration as per Monday 10th proposal. …I then thought it would be better to let them go quietly to work than execute ye justices orders any further than telling what would have been the case had they persisted in their foolishness. They all seemed convinced of their errors and only desired a little matter to drink. Mr Russell and Mr Wade (the owners) being present, agreed to allow them a half bottle of ale' which they drank very quietly.'

The following day the men were all back to work. However, trouble broke out again over the size of the corves, the hazel baskets used to carry the coal out of the mine. On the 28th July 1780, Gibson recorded that:

> 'The pits are all idle to measure the corves – they having tried some yesterday and said they were too big. I had a deal of altercation with the men and upon measuring found ye Corves too little. After a deal of inconsistent altercation they all resolved to go to work.'

At this time Gibson was also busy at Wallsend Colliery, belonging to the Chapman brothers, where Russell and Wade also had an interest. On the 17th May 1780, he 'met Mr Brown and Mr Chapman and designed the Waggon Way a little'. At the beginning of July, travelled to Hexham and to a wood nearby where 'he marked 50 pieces of Timber for Wallsend Staith'. On Saturday 29th July, he recorded reaching coal at a depth of 101.5 fathoms which was followed by a celebration. He noted 'we had a glass (rather too much on ye occasion) and got home latish'. The next day 'I was at home (not well) till noon then went to Whitburn and Bathed'. On Monday 27th November, he recorded that 'the Spout is now finished and we had two waggons of coal sent down'. This marked the beginning of the shipment of the prized household coal from the High Main seam at Wallsend Colliery, which was to become a byword for quality throughout the coalfield. Sadly, William Gibson was killed in an accident at Washington Colliery in 1783 before Wallsend's reputation was firmly established.

Such was the life of the viewer, that multi skilled engineer and diplomat at the centre of all the operations of the colliery. These were the men who had the professional knowledge to win the coal and the personal skills to manage the workforce and the owners. The next chapter considers William Brown's management of Throckley and Walbottle collieries to illustrate the nature of coalmining and the life of the miners in the eighteenth century.

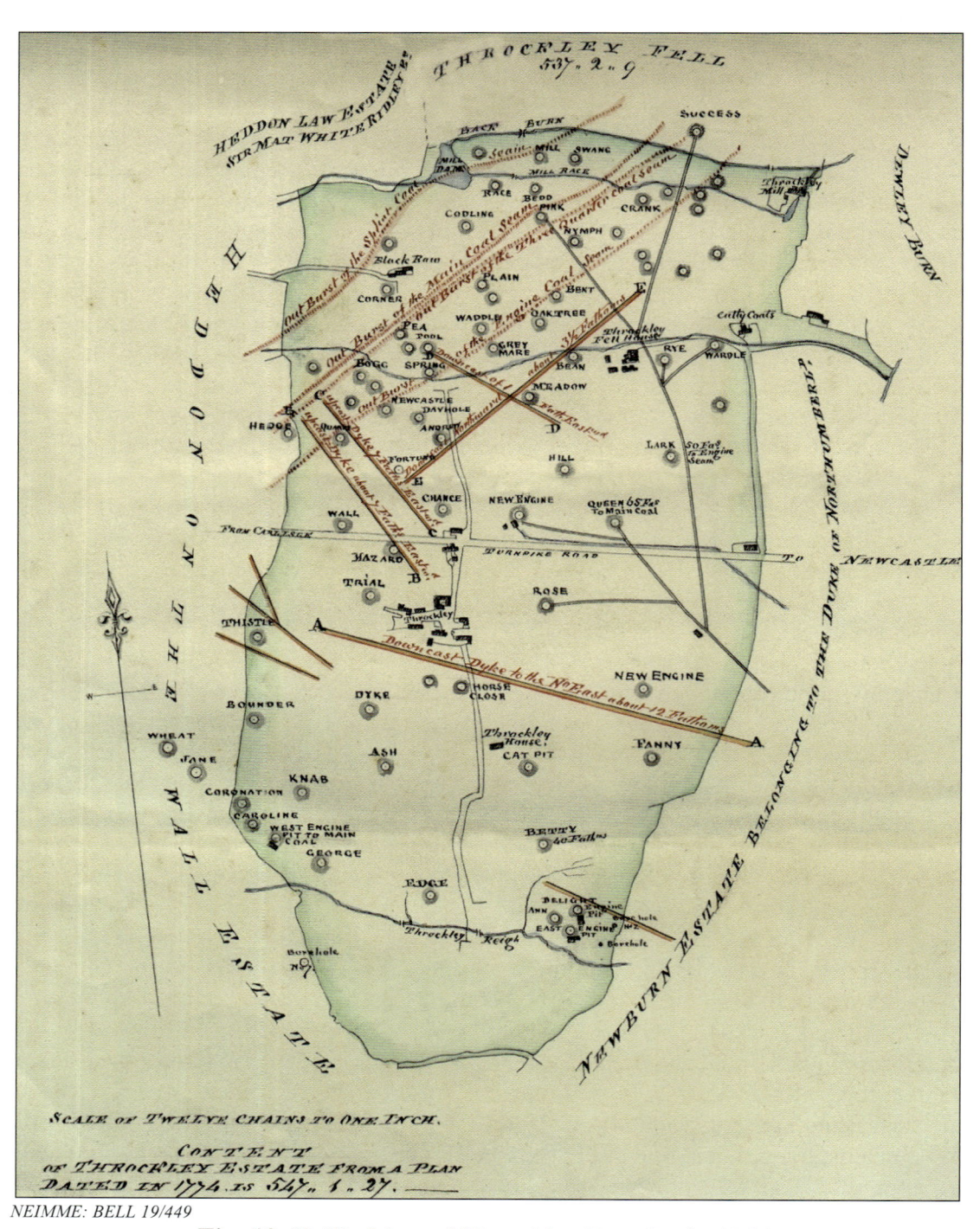

NEIMME: BELL 19/449

Fig. 13: Bell's Map of Throckley Royalty in 1781

Chapter Five: Throckley and Walbottle Collieries

Throckley royalty had been owned by the Earls of Derwentwater since the fifteenth century but it was seized by the Crown in 1716 as punishment for the third earl's involvement in the Jacobite rebellion. In 1735, King George I granted the estate to the Royal Naval Hospital for Seamen, better known as Greenwich Hospital. A rental of that year lists William Brown Senior as the tenant of Throckley Colliery. Upon his death in 1746 the colliery was managed by his son. At that time Throckley was a landsale colliery but this was nonetheless a sizeable business: in the season 1750-1 it produced 10,000 fothers (8,750 tons), a fother being a cartload weighing about 17.5 cwt.[78] However, the major profits were to be made in the seasale trade with London and in about 1749 a partnership was formed between William Brown, Henry Masterman, William Barkas and Mr Alcock to develop Throckley as a seasale colliery. William Brown Junior was the engineer responsible for the development of the seasale colliery; Henry Masterman, the owner of Newburn Hall, was an attorney based in London; William Barkas, who had been enrolled as a member of the Hostman's Company in 1749, was the fitter responsible for marketing the coal; and the solicitor Mr Alcock, who had offices in the Custom's House in Newcastle, seems to have been in charge of the accounts. Each member held a quarter of the shares. The partnership did not always run smoothly for there were tensions between Masterman and Barkas. In a letter to William Brown dated 10th May 1753, Henry Masterman commented that 'I have such an opinion of Barkas that I would not trust him out of my sight and therefore I think myself extremely happy in having two such friends as you and Mr Alcock to look after him for me'.[79] Masterman asked William Brown to send him regular reports on the operation of the business.

Throckley estate was 547 acres in area and there were four seams of marketable coal. Bell's map (fig.13) shows the position of the outcrops of the principal seams which are referred to rather dramatically as 'bursts'. It also shows the major geological faults. A view of Throckley Colliery made by George Johnson in 1784 reported that the best coal for the London and coast markets – the seasale trade – was the Splint Coal, the deepest seam. This was also the opinion of George Green, the viewer of Heddon Colliery, who ranked the Engine seam as second best, followed by the Three Quarter and the Main. His report is summarized below.[80]

Seam	Depth	Height	Rank
Engine	36F from surface	3'6" high	ranked second
Three Quarter	48F from surface	2'3" high	ranked third
Main	50.5F from surface	3'6" high	ranked fourth
Splint	68 F from surface	3'6" high	ranked first

In 1774 William Brown provided an estimate of the coal remaining in these four seams. From a total of 547 acres, only 130 acres remained in the Engine seam where most of the earlier mining activity had taken place; 400 acres remained in the Three Quarter and Main seams; but the best seam, the deep Splint coal, was largely intact with 500 acres

---

[78] Northumberland Record Office: ZCK 14/2. NEIMME: Brown/I/9 William Brown to Carlisle Spedding, 12th July 1750.

[79] NEIMME: Brown/I/278, Henry Masterman to William Brown, 10th May 1753.

[80] NEIMME: Watson 2/8/21 and 184. Details of the geology of individual pits are recorded in the book 'Account of the Strata of Northumberland and Durham' published by NEIMME in 1885. Much of this information is also available on the website of Durham Mining Museum.

remaining. There are over seventy pits, spaced close together, marked on Bell's map (fig.13) and this was typical of collieries of the period. The mines were generally between 20 and 50 fathoms deep and operated in an area about 200 yards from the shaft which was the limit of the ventilation system of the day. Generally, a pit was in production for a period of about four to ten years. The seasale pits were served by branches from the waggonway but the route of the branches frequently changed: when one pit closed the railway was lifted and moved to another pit about to be opened. Thus, the waggonway shown on Bell's map only illustrates the section which was in operation in 1781.

In a report dated 1774, William Brown described the principal stages in the development of Throckley Colliery.[81] He began by noting that 'most probably the First Winning of the Engine Seam was out of Heddon Banks by a Water Engine Wrought by the Tyne about 110 years ago' (c.1664). This would be a waterwheel driving a battery of pumps often referred to as a water mill. 'The 2nd Winning was about 80 years ago by a level drove out of Newburn Deen to the South East side of the Moor' (c.1694). This was probably what was later referred to as the Tyne Level which ran due east to west through Throckley and Heddon in a position just north of the site of the later pumping engines.[82] The level enabled the coal from the Engine seam near the outcrops to be won. This was the seam which the Brown family were mining in the 1730s, 1740s and 1750s. The next major developments were all associated with the construction of pumping engines in about 1760, 1765 and 1767. Brown noted that 'the Colliery was also won about 15 years ago (c.1759) by a Fire Engine Built on the South West side of Throckley Bank'. The date of the building of the west engine is recorded in the names given to the pits – Coronation, George and Caroline – commemorating the accession of King George III in 1760. However, Brown commented that 'the Effectual Winning was about 9 years ago by a powerful Fire Engine erected in the South East Dip part of the estate near Newburn Grindstones by which the colliery has been wrought ever since'. Brown was referring to the Delight Engine built in 1765. He added that 'A third Engine was about that time erected near the last mentioned one and a pit sunk to the main coal there at 15 Fathoms below the Engine Seam but the main coal did not answer to Expectation therefore that Engine is discontinued at present'. The final stage was when 'A Fire Engine was lately erected on the Northside of the said dyke' which would 'win the remainder of the Engine Seam being about 100 acres'. This fourth engine was north of the turnpike road and it is shown on a map of the estate dated 1769. A 'new engine pit' is also marked further south indicating that a fifth engine was planned but was not installed by 1774. All five engines are marked on the 1781 map (fig.13).[83] This overview is helpful in so far as it provides a framework for understanding William Brown's correspondence. At the time of Brown's report, 1774, major changes were afoot at Throckley Colliery. The West Engine and the Delight Engine were being decommissioned and recycled. Many of the parts were used to build the great engine at Willington Colliery where the partnership had taken out leases in 1773. The remainder was used to build the sixth engine on Throckley Fell, north of Dewley Mill, shown in fig.14. Unfortunately, the branches of the waggonway are not marked on this later map.

---

[81] NEIMME: Watson 2/5/217
[82] NEIMME: Watson 3/113/4
[83] NRO: 0536-01 A Plan of the Inclosed Lands in the Lordship of Throckley 1769. Fig. 13 is a plan from the Bell Collection and it is part of a series commissioned in the nineteenth century to illustrate aspects of mining in the area. It is based upon Watson 23a/21 which dates to 1781. The reference to a plan of 1774, now lost, refers to the acreage – the content.

There was usually a considerable time gap between signing the lease and selling the first coals, a time when there was a large outlay of capital before any returns were forthcoming. By autumn 1751, William Brown had the colliery ready to supply the seasale trade and informed his friend Carlisle Spedding in November that

> 'I have got our colliery at Throckley in a pretty fair way having had a new Coale pitt at Coale work a Month and has another Coaled this day I propose to work a Pritty Large quantity of Coales out of the 2nd pitt (viz) in that I have at work, I raise my water 7 fathoms and works 12 or 18 xx of 16 Peck Corves p. Day and will in 20 Days time be able to work in the new one nothing less than 18 or 20 xx p. day. She is 39 and the one at work 49 fathoms deep. We have also got our waggon way completed and Staith so far forwd as to Load Keels at the Spout and Indeed has loaded 4 Large Ships since we began to lead. We have still high thought they will be at the old price viz 13s p. Chalder'.[84]

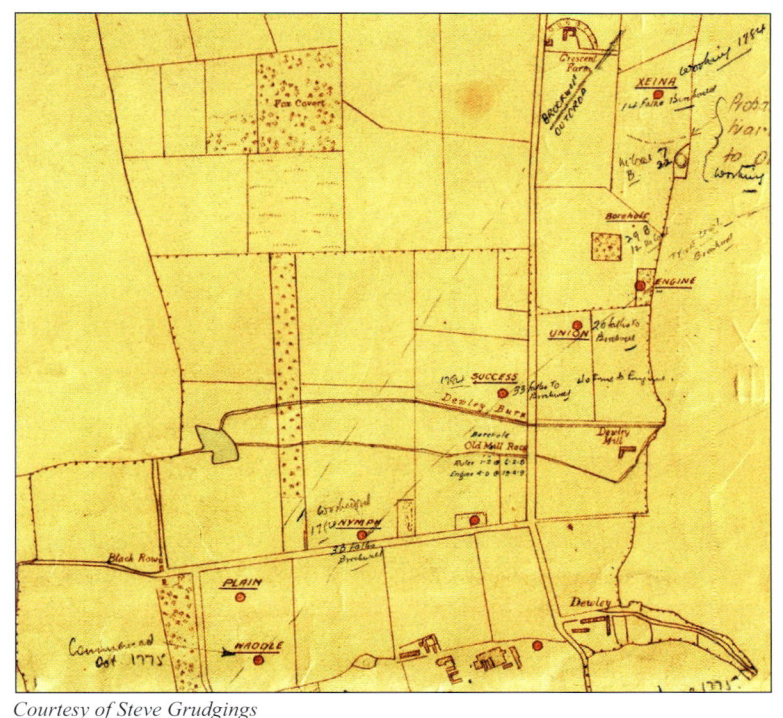

Courtesy of Steve Grudgings

Fig. 14: Mining on Throckley Fell in 1785

In January 1752, William Brown reported to Henry Masterman that 'will be able to work tolerable quantity if can get men for her and am under no fear but Mr Barcas will vend them for has tried them at several Markets along the coast as well as Yours at London'. He added that 'we are very farr with the Roofing of our staith and believe in 20 days time will have it completed. It is the most commodious one in the River and can load coals at anytime'.[85] The engraving of Holiwell Reins staith (fig.15) shows the type of large wooden structure built to protect the coal from weathering, which reduced its price on the market. A loaded keel boat prepares to move upstream on the ebb tide with its cargo for the sea-going vessels – the colliers – moored at Shields, a journey of some fifteen miles. Other keels are tied up alongside the staith.

---

[84] NEIMME: Brown/I/35 William Brown to Carlisle Spedding, 2nd November 1751. 'xx' means score.

[85] NEIMME: Brown/I/46 William Brown to Henry Masterman, 10th January 1752.

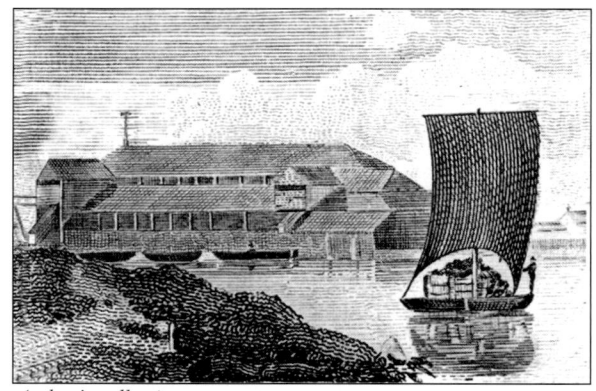

Author's collection
Fig. 15: Holiwell Reins Staith at Lemington

In the early 1750's, the pits were situated west and north of the medieval village. The building of General Wade's Road, the Newcastle to Carlisle turnpike, in 1751 was a major benefit to the landsale trade since it provided much better access to markets both to the east and west. The seasale trade was served by a waggonway, built in the same year. The branch from the Chance Pit was joined by a branch from the Thistle and Trial pits just east of the village. The main way ran eastwards from the village on the south side of the Military Road before turning south along the western edge of Walbottle Dene. The ravine was crossed on a large wooden bridge en route downhill to Lemington (fig.17).[86] However, there was trouble from their neighbour and competitor in the trade, Mr. Humble, who was mining near Pigg's Hall on Newburn Moor. Brown's letter to Masterman reveals something of the cut throat nature of the trade at this time.

> 'As to our waggon way it stands Extraordinarily well when Mallicious people letts it alone but on the 3rd January 8 or 9 men by Mr Jno Humble's order tore up a considerable piece in a Close belonging to him called Jolleys Close not withstanding he had let the said close to Mr Barcas for the Term of Nine years in respect of all minerals…..also the same day, the same people, and by the same order, tore up another piece in a field of yours called Low Hill Nigh Lemington and threaten and do more mischief upon which our waggons was stopped.'[87]

Humble was taken to court and the waggonway re-opened but Henry Masterman expressed his concern for the future in his reply to William Brown.

> 'What to be done with Humble with regard to wayleave thro' Close at the end of 9 yrs, tho' it is to be hoped by that time we shall be able to lay out a new way in case he should refuse to enlarge the term.'[88]

This is a good example of the situation faced by many waggonway engineers who had to deal with the problem of re-routing the way to overcome difficulties with wayleave rights. Despite these problems, Brown informed Spedding that Throckley Colliery was 'very good last season'.[89]

---

[86] The following pits are mentioned in the correspondence dated 1749 - 55: Tryall, Hazard, Chance, Fortune, Industry, Dayhole, Newcastle and the Thistle which was also linked to Heddon Colliery. The line of the waggonway is shown on a map dated 1755 ref. Alnwick Castle/0/17/4/2 Letter from W. Brown to Messrs Thynne, Seymour and Scott, 17/6/1755.
[87] NEIMME: Brown/I/46 William Brown to Henry Masterman, 10th January 1752.
[88] NEIMME: Brown/I/49 Henry Masterman to William Brown, 21st January 1752.
[89] NEIMME: Brown/I/37 William Brown to Carlysle Spedding, 9th March 1752.

Part of William Brown's job as viewer was to plan the development of the colliery and organise the sinking of pits; he also had to recruit, manage and retain the labour force needed to maintain production. In 1752, he informed Masterman that 'we have begun the Pittman's houses and hopes to have 12 up by 1st May'. The provision of good accommodation was an important factor in recruiting and retaining men.[90] Brown also discussed a more ambitious plan and Masterman replied enthusiastically 'I shou'd think that getting us a Fire Engine at once and making a winning at the foot of the Banks will be the best thing we can possibly do'.[91] Brown shared his thoughts with Carlisle Spedding:

> 'Thought of fixing a fire engine six hundred yards to the dip of our Tyne Level by virtue of which we now work and if that is done will be able to work upwards of 25,000 Chaldrons each year for the Term of our Lease for our Boundaries is great length. Two of my partners is at London at this juncture Consulting with the Third who lives at that metropolis about that affair and if its resolv'd on to Take a Trip to your Place in order to begg a Little of your advice as to the erecting of one of these machines.'[92]

A report by William Newton, Amos Barnes and George Claughton recommended that the 'Cylinder of the Fire Engine…be no less than Forty-Two Inches Diameter' and noted that 'when Completed will at Least win a Hundred Acres of Coal in the South West Side of the great Dyke'. They estimated the cost of erecting the Fire Engine, Sinking the Engine Pit and one Coal Pit' to be £1,100.[93] By November 1752, Brown had sunk the Engine Pit six fathoms and was in negotiation with Isaac Thompson of the Coalbrookdale Company for an engine. By the following summer, Brown was able to report that 'we have made great progress with our intended Fire Engine'.[94] However, Barkas was not supportive and nine months later wrote that 'he cannot at present bring himself to think that a Fire Engine is necessary'.[95] Brown shelved the idea for another reason: 'I had a very good Vend of Coales from our Colly which promises a good deal better than she did some time ago So that we have no Imediate thoughts of Erecting a Fire Engine'. It would be another six years before an engine was built.[96]

In the mid eighteenth century, the map maker Isaac Thompson was an influential member of northern society, being the proprietor of the Newcastle Journal, which he had founded in 1739. He was a Quaker by religion and a man of many talents who was noted for his interest in science and philosophy. He worked as a land agent and surveyor but was also a printer, publisher and newspaper proprietor.[97] His maps, such as the plan of East Denton and the survey of the Duke of Northumberland's Newburn estate, are an important treasure trove of information relating to the coal industry. He was related by marriage to Abraham Darby and this relationship, together with his interest in science, led to his appointment as the representative in the North East of the famous Coalbrookdale foundry in Shropshire, which supplied cylinders and iron pipes for pumping engines; and in this capacity he had close links with William Brown. Whether

---

[90] NEIMME: Brown/1/55 William Brown to Henry Masterman, 7th April 1752.
[91] NEIMME: Brown/I/76 Henry Masterman to William Brown, 11th August 1752.
[92] NEIMME: Brown/I/68 William Brown to Carlyle Spedding, 11th July 1752.
   Wat. 3/113/3 The Tyne level ran east west and drained into the New Burn.
[93] NEIMME: Forster 1/4/231.
[94] NEIMME: Brown/I/249 William Brown to Carlyle Spedding, 15th July 1753.
[95] NEIMME: Brown/I/285 Henry Masterman to William Brown, 14th March 1754.
[96] NEIMME: Brown/I/159 William Brown – Leonard Hartley 10th May 1754
[97] R. Welford, Early Newcastle Typography 1639 – 1800, Archaeologia Aeliana 3rd Ser

Isaac Thompson had any connection with the foundry of Jonathan Thompson in Gateshead, which also supplied iron work to the coal trade, is not known.

In 1753, William Brown was able to report to Henry Masterman that 'we have paid off most of our debt so that I hope this year will give you the satisfaction you have so long expected'. However, there were problems arising from newcomers into the market and Brown suggested an interesting solution which illustrates the power of wayleave rights in the politics of coalmining.

> 'The new colliery at Wylam setting forth…will affect us in our Landsale for at present nobody interferes with us to the west but I am sure it will affect us more in inveigling our workmen from us (we have now a good set of hands) and they have begun something of that kind already; therefore as their owners must unavoidably petition you for wayleave to lead their Coals through your Lands at Newburn Hall, I apprehend it would be greatly to the advantage of Throckley Colly if you wd be pleased to restrain them from employing any men or lads that has belonged or wrought with us any time for two years past or may happen to work with us hereafter; and it would be well if they were restrained from selling any Coals by Landsale. The same thing may happen to us from Mr Beaumont's Colliery at Newbiggin and Whorlton Moor. If they succeed these Collys, as well as Wylam, must come through your lands at Newburn Hall and I am concern'd must submit to such terms as you please to grant them.'[98]

The Earl of Northumberland had claimed a right to the coal in Throckley and Masterman in his response to the problem posed by Blackett and Beaumont was anxious not to alienate the lord of the manor of Newburn at a time when the partnership was seeking to extend its waggonway lease.[99]

> 'We must manage our affairs with so much prudence and Discretion as to avoid giving Ld Northumberland any Displeasure of Disgust because we don't know what Connexions we may hereafter happen to have with his Lordship.'

The following autumn, the time when the miners were required to sign the annual bond, William Brown's fears were realised. He bluntly reminded Masterman of the consequences of not heeding his warning: the binding money paid to attract the miners to sign for the partnership had gone up by £150 and he was forced to travel to Scotland to seek men.

> 'I have had a troublesome and expensive piece of work to get our men bound mostly sanctioned by our Neighbouring Collierys going on. Mr Beaumont sent his servants among our men to their very habitations, to treat with them and in spite of all I could do has got two of our best hands and Mr Blackett has got one, in order to keep what men we had gave them an advance Price which will be £150 a year upon the Colly and if our people had not been tempt by our good Neighbour I believe I could have reduced the prices. In a letter of mine some time ago I gave you a hint of what I feared would happen which hint you took little notice of tho' was persuaded you had it in your power to have made these Gentlemen behave in an other manner. I have been in Scotland 20 miles beyond

---

[98] NEIMME: Brown/I/282 William Brown to Henry Masterman, 26th February 1754.
[99] NEIMME: Brown/I/285 Henry Masterman to William Brown, 14th March 1754

> Edinburgh seeking some hands and have already had some from that part of the World and am sure will get several more. Was at 6 collieries in Scotland and 3 near Berwick wh cost me much money and trouble. I will in all probability be in a capacity to work 20,000 Newcastle Chalders next year if Barkas can but vend them. I hope to clear most of the debt the Collys in.'[100]

At a time when the coal industry was expanding, all colliery viewers continually faced the problem of securing skilled labour. In February 1755, William Brown received a letter from his business associate John Burrell, a leasee of the Duke of Hamilton's colliery at Bo'ness, which gives an insight into the extraordinary efforts which coalowners were compelled to make.

> 'In my last jaunt to the west I was not good in procuring pickmen both for you and for me. Those who have families I propose to take and the yonkers who have nothing to care for but themselves to send to you…it is the Colliers about Glasgow I have the most dependence upon as they seem to be or at least will soon be in a Staggering Condition. So soon as those I have be spoke come to hand I shall afford them some little money to carry them your length'.[101]

By the following summer, John Blackett was on the point of winning Wylam Colliery and at the October binding time Brown wrote to Masterman about 'the fighting wch I forsaw wd ensue between Mr Blackett and Mr Beaumont owner of Newbiggin Colly and the Colliery at Throckley at Binding the pittmen' which had cost the partnership £200 or more in additional binding money. To put this figure in some context, it amounted to the total profit from the successful landsale colliery at Throckley in the 1753-4 season.

> 'The truth is that Mr Blackett or his Agent has been inveighing our Men and has made them so thicklish, and they being naturally given to ramble, I declare I do not know upon what Terms I will be able to hire them and am doubtful unless you give these gentlemen (especially Mr Blackett) a Charge I will loose servants. If you chose to say any thing the sooner the better.'[102]

Once again, Masterman refused to take positive action reminding his partner that 'it is the practice of all Collierys and I don't see how tis possible to prevent it'. He also complained about the shortfall of Throckley coals on the London market and the men's enjoyment of horse racing commenting that 'these things I am convinced have contributed more to the debauching the minds of your workmen than any one can imagine'.[103] In his reply Brown avoided mentioning the races but reassured Masterman about the future.

> 'The steath has not been clear'd of Coals this season except 5 or 6 days in ye middle of October and that was ye time our people wd not work full work expecting to bring in better terms thereby for the present year. I have however got a great Number of Men bound and upon rather better terms than last year

---

[100] NEIMME: Brown/I/302, William Brown to Henry Masterman, 30th November 1754.
[101] NEIMME: Brown/I/242, John Burrell to William Brown, 17th February 1755.
[102] NEIMME: Brown/I/319, William Brown to Henry Masterman, 17th October 1755.
[103] NEIMME: Brown/I/320, Henry Masterman to William Brown, 1st November 1755.

> and will ye ensuing year to able to work a quantity of Coals equal to most of the Considerable Collys in ye River.'[104]

In 1753, the partnership had leased Heddon Colliery from the Earl of Carlisle and by the beginning of 1754 there were two pits at work. Throckley and Heddon collieries were operated as a single enterprise which was involved in all three aspects of the coal trade – landsale, seasale and export overseas. On 17th June 1755, Brown wrote to the Duke of Northumberland requesting an extension of the lease for the waggonway through Newburn.[105] Brown was also seeking a realignment of the route through the Duke's lands in Newburn to accommodate a new branch from the Queen Pit of Throckley Colliery in the north and another branch from the southern part of Heddon Colliery in the south west. On the accompanying map (fig.21), the Heddon branch is marked running in a north, north easterly direction, passing north of Throckley House, to join the line from the Queen Pit before heading towards the bridge across Walbottle Dean. The route still exists today as a farm track.

In March 1756, Brown announced that four new coal pits had been opened.

> 'I have at present four working pitts out of which I can work a great quantity of Coales Round and Dry. In fact I can work more than 20,000 Newcastle Chalder year for years to come without being of much more expence of sinking and has got the waggon ways laid and completed to the respective new pits so that now if you can sell I can supply you. The Coale in the new pitts which is in another part of the Estate.'[106]

The four pits were probably the Knab, Ash, Dyke and Bounder, which were working the Engine seam, and they were probably linked by branch lines to the existing waggonway from Heddon Colliery.

The first fire engine was placed in the south west corner of the estate and drained down to the Main seam. It is marked as 'West Engine' in Fig.13. The names of the pits, George, Caroline and Coronation date the development to 1760, not 1756 as is usually stated. In 1765 and 1766, another two engines were added further east and William Brown reported to James Spedding, Carlisle's son, that 'in the spring last year I began to Win Throckley Colliery at the Extreme Deep', the south east corner. The Betty, Delight, Catt and Edge pits were opened between 1765 and 1768.[107] All these pits were linked to branch line which joined the Wylam Way at the washhole in Throckley Reigh.

After sinking the shaft to win the coal, William Brown made a contract with individual managers to work the coal: in October 1765, for example, he signed a contract with William and John Robson to 'do all manner of work underground ... for carrying on the Betty Pit in Throckley Colliery and Engine Seam'. This included running 'each Board from the Main Headways 200 yards in a straight line': the distance probably represented the limitations of the ventilation system of the day. Headways were driven from the bottom of the shaft to open up the colliery and the bords, where the hewers extracted the coal, were worked at right angles to enable the men to hew across the

---

[104] NEIMME: Brown/I/322, William Brown to Henry Masterman, 13th December 1755.
[105] Alnwick Castle/0/17/4/2 Letter from W. Brown to Messrs Thynne, Seymour and Scott, 17/6/1755.
[106] NEIMME: Brown/I/225 William Brown to Cpt John Biggins, 20th March 1756.
[107] NEIMME: Brown/I/38 William Brown to James Spedding, April 12th 1765; Watson 2/12/3,19,21 and 77; Watson 2/13/5 an agreement with Nicholas Chance to build the Engine House at Delight Pit.

grain of the coal. The Robsons also had to hole the Walls at every twenty yards: that is to drive a passage to link the different boards thus forming a pillar to support the roof. William Brown would find 'men, lads and Horses for carrying on the said work' and to pay them; but contractors were required to find all other labour and 'oyl, candles, lay all barroways, set all props, having wood found him what is necessary'. The rates for this work were based upon the output of the pit: 6d per score for every score put with trams and 10d for every score put with horses 'which sum the said Wm Brown agrees to Pay once in every 14 days'. Each pit would mine about fifteen acres of coal.[108]

Courtesy of the Society of Antiquaries of Newcastle upon Tyne: NRO: SANT-BEQ-05-03-14-36
Fig. 16: Walbottle in 1767

In 1767, when Isaac Thompson produced his plan of the Duke of Northumberland's Newburn estate (fig.16), the colliery was on the verge of a major development under the direction of William Brown.[109] Humble's colliery near Lemington had ceased production a decade earlier and the new centre of activity in Newburn royalty was near the village of Walbottle further west. The three exploratory boreholes for Walbottle Colliery, G, F and H, are marked to the west of the village on Thompson's plan. The colliery accounts for the period November 1768 to November 1770 provide an unusual amount of invaluable detail about the men employed in sinking the Engine and Betty pits. Barnabus Brough was the sinker, George Penman the mason, William Puncheon the carpenter, John Hall the smith, Richard Bevans supplied nails and Thomas Falcus supplied '26 suites of Flanning Cloaths for the Sinkers'. In November 1769, the sinkers were paid a gratuity of ten guineas 'for their good Behaviour and getting the Coal within a year from the first Beginning to Sink'. The stages by which the engine shaft was sunk, a drainage level was driven, a reservoir and aqueduct was constructed to supply water to the engine, the engine house was built with two dwelling rooms, and the parts of the engine were assembled and erected, including the purchase of a second hand boiler, 14 foot in diameter, from Alderman Ridley and a 70-inch cylinder – all are recorded.[110]

---

[108] NEIMME: Watson 2/13/19
[109] Northumberland Record Office: Sant/Beq/05/03/14/36
[110] NEIMME: Watson 2/7/139 – 260. Watson 2/7/15 gives a detailed costing of the Walbottle Engine

William Brown's plan of 1769 (fig.17) shows Walbottle Colliery in the early stages of its development. It is particularly interesting to note the reservoir and aqueduct built to provide a constant supply of water to the fire engine.

Some understanding of the workforce needed to operate a major colliery, such as Walbottle, in the late eighteenth century is provided by a document in the Montagu papers.[111] Edward Montagu was given an assessment of the key workers needed to open a pit at East Denton which was probably compiled by the viewer William Newton, who had served the previous owner of the estate, John Rogers. The document records that 'An Agent is necessary that is well acquainted with Collierys as a viewer who is to View the Colliery & look to the Engine & Staith also & keep the Accounts with a Clark for his Assistance'. Besides the viewer, there were three levels of supervisors at the colliery – the overmen, deputies and keekers. Montagu was advised that one overman 'has the Management of the Pit Crew & directing the working of the Coals so as to make of them as dry & as marchantible as possible'. Deputies and keekers served as lesser officials under the overman.

Montagu was made aware of the importance of securing a good 'Engine Wright to take care of the Engine & two men to conduit it, the one of them to be a Blacksmith'. Not surprisingly because of the crucial importance of the pumping engine at many collieries, these men were amongst the highest paid employees. 'A Staithman is also necessary to have Charge of Receiving the Coal from the Waggons & Delivering the same to the Fitters. This Staithman takes care to have the Waggons or other Carriages to bring him such Quantities of Coals to the Staith as are agreed upon & to give the Keelmen their proper measure & to keep account of everything relating to the leading & Delivery of the Coals and also of laying them properly on the Staith when the Keels do not take them away as led'. Also, 'there must be a Fitter Employed to sell the Coals' who was paid by the volume of coal handled. The other workmen needed were 'the Banksman, the Corver, the Waggon Way keeper & Waggon Wright & Blacksmith, Horsekeeper, Inspectors at the pits, Waylers etc'. The banksman was responsible for the colliery above ground: he off-loaded the corves at the pit head, dispatched them to the screens to remove the stone and small coals and marshalled the chaldron waggons for the railway (fig.9). The corver was the man responsible for making the wicker baskets used throughout the mine to carry the coal from the face to the surface. This was a skilled job because the size of corve was used as a measurement to determined the volume of coal extracted by the hewer and therefore his pay. The waggon way keeper and the waggon wright were responsible for maintaining the railway to the staiths. The blacksmith would be needed to assist the railwaymen but his principal role was maintaining the pumping engine and sharpening the picks of the hewers. Waylers were used to sort stone from the coal at the screens and this was the only job open to women who were not employed underground in the Great Northern Coalfield. In addition, there were the hewers, barrowmen, or putters, and their boys to employ and this last group of workers constituted the bulk of the labour force.

---

[111] NRO: Sant/Gen/Est/1/4/2/10

William Brown's Plan of Walbottle Dene 1769

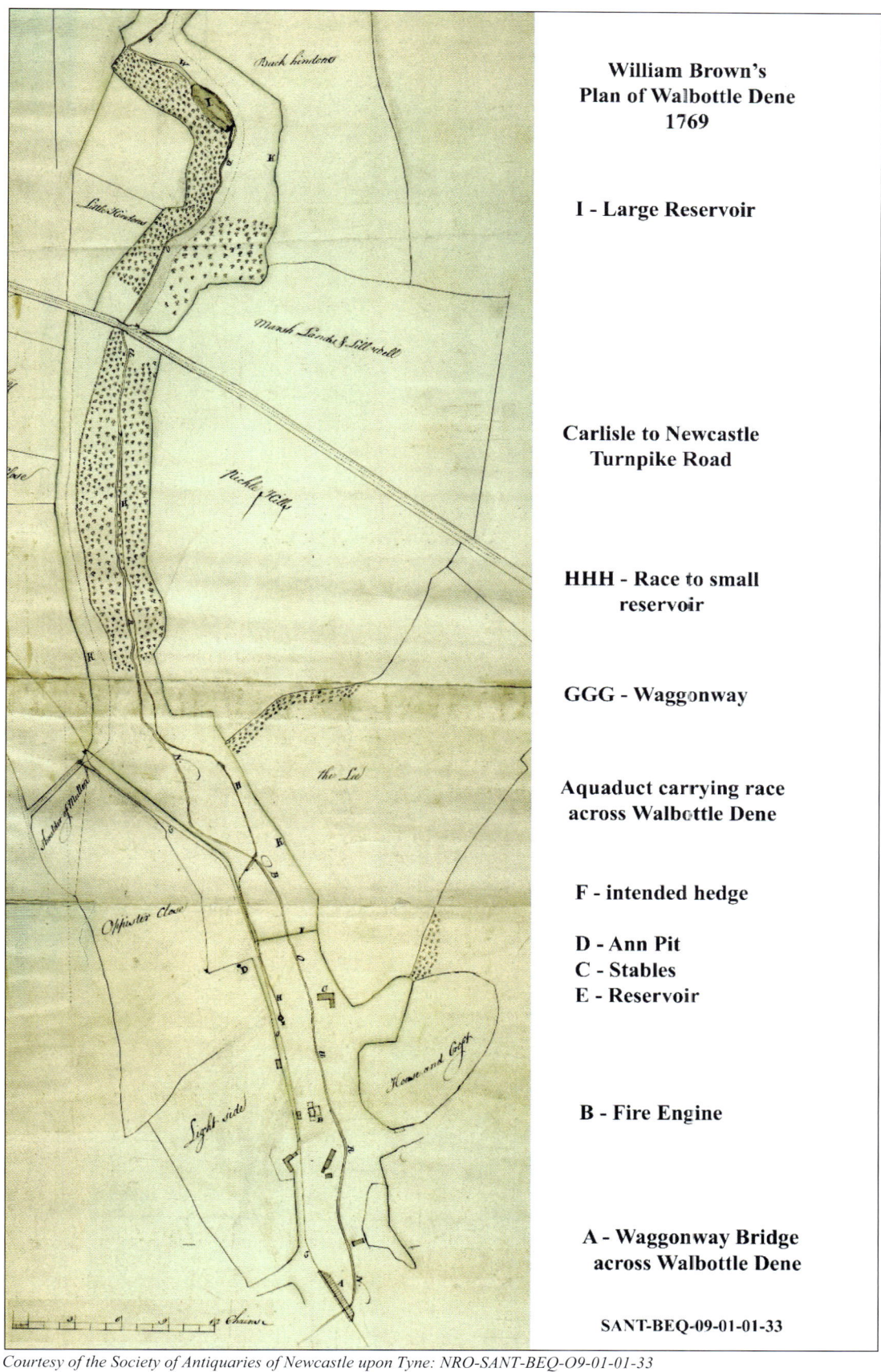

I - Large Reservoir

Carlisle to Newcastle Turnpike Road

HHH - Race to small reservoir

GGG - Waggonway

Aquaduct carrying race across Walbottle Dene

F - intended hedge

D - Ann Pit
C - Stables
E - Reservoir

B - Fire Engine

A - Waggonway Bridge across Walbottle Dene

SANT-BEQ-09-01-01-33

Courtesy of the Society of Antiquaries of Newcastle upon Tyne: NRO-SANT-BEQ-O9-01-01-33

Fig. 17: Walbottle Dene in 1769

In November 1769, hewers were bound to 'work in the Engine seam at 2/- per score when working in the whole mine and 1/10 per score when working pillars; a score to be 21 corves of coal and each to contain sixteen pecks'. The men were required to send 'to Bank good, clean and Merchantable Ship Coals'. The putters or barrowmen were required 'to put the coals at 8d per score until the workings are 60 yards from the shaft and the sum of 9d per score while the workings are between 60 and 80 yards from the Pitt or Shaft'. The drivers 'shall drive the horses underground at the Rate or Price of 1/- p.day'. Interestingly, the men were required to be flexible: 'those that are hewers and those that are Barrowmen or Drivers of Horses Underground shall and will in case of Necessity…do the business of each other'. All men were required 'not at any time to absent themselves from the work…or to work at any other colliery or any other work whatsoever without the License and Consent of the said Duke of Northumberland….unless sufficient cause can be shown for such absence'. The penalty was '1/- per day plus a penal sum of £50'. This amounted to much more than a year's wages and was indeed penal. The Duke 'shall provide such candles or lamps as shall be by them consum'd in the said Colliery underground in putting Coals but not in Hewing them'. In addition, he would 'provide all materials – deals, props, trams, sledges, shovels, mauls and wedges and shall cause the picks to be sharp'd and carry'd but not to find and provide picks'. It is interesting to note the Duke's concern for the welfare of the workforce in the event of an accident occurring: 'If it shall at any time so happen that any Casualty or disaster of any kind whatsoever shall befall the said Colliery….so that the said parties shall be hindered from working therein'…the Duke would 'provide alternative work and pay the said parties fair and sufficient wages for such Labour'.[112]

The opening of any colliery was the occasion for a party. Walbottle Colliery accounts for the fortnight April 4th – April 13th 1770 record payment 'for a Treat given to the Waggon Men, Keelmen and Workmen at the first day the Waggons came down'.[113] No detailed account survives of the celebrations at Walbottle but Elizabeth Montagu gives a graphic description of the events at the neighbouring colliery of East Denton in a letter to her sister and it is safe to assume that something similar took place.

> 'The black Gentlemen, after much necessary ablution in the Tyne, put on clean linen and their holiday cloaths, then they assembled at the Engine house from whence they proceed in manner following. First marches a Flag fluttering in the wind, then follows a French horn, then the respectable corfe of coal carried by 2 men, on each side a Fiddler, then next to the corfe in place of honour, the Engine men and master marching two and two, then the staithmen and overmen, then the Pittmen, all two and two, thus they march, musick playing, colours flying, in most orderly procession. When they came to the place opposite the gate (of East Denton Hall) they marched quite round, when came a second time to ye gate they set down the corfe and gave a most notable shout. We went to ye gate, thanked them for their good wishes, and told them we had provided a dinner for them. They gave us three cheers one of which would have made ye headack for a week. Then most solemnly they paced to a large field opposite the garden, in which we had made expansive tables and benches, a coal owner being ever provided with deal boards in abundance. The guests invited were 377, the smell of ye feast attracted more. Here they were served with a large fat sheep roasted whole, another boiled in joints, a vast mass of beef, seven large pies and

---

[112] NEIMME: Watson 2/12/109.
[113] NEIMME: Watson 2/7/264.

twelve plumb puddings longer than the diameter of the moons orbit and as much ale and punch as they wd drink, which was indeed as much as they could drink, the music playing ye whole time. As soon as they had dined we went to them. The moment we appeared they stood up and shouted. We went first to the table where the tenants and the head work men sat, hoped they had had a good dinner and bad them welcome. Then to ye Pittmens, then to the women's table, abundant thanks we received from all. Then we went again to the first table and drank their healths and success to the Colliery, when they all gave three cheers.'

This was quite a party: Mrs Montagu noted that 'at twelve those who could, departed' but 'alas, many lay there all night…drunk upon the field of battle, others in the dry ditches by the way side'[114].

Mrs Montagu's viewer, William Thomas, in response to an exhaustive inquiry by Sir John Swinburne into the living and working conditions of miners and their families on Tyneside in 1800 provides some fascinating detail about the mining communities.[115] Thomas contrasts the high wages and high spending of the colliers with that of the farm labourers noting that 'Intemperance is a leading feature in the character of a pitman' who indulged in ale to excess rather than spirits. Cockfighting was their favourite sport and 'they often bet considerable sums on these matches which seldom conclude without quarrelling and often end in blows'. In the management's view, high wages were the source of the 'Vices which attach to the character of the pitman' since they are 'more than sufficient to provide for the necessities of life' and 'the surplus is dissipated in Riot and Intemperance'. He regarded the wives of pitmen with particular scorn being 'a very indolent set of women' and 'strangers to cleanliness, frugality or economy'. They were slovenly in their persons and their houses. Thomas lamented that 'little attention is paid to the duties of religion'. Despite Elizabeth Montagu providing a school, the formal education of their children was not a matter which the colliers took seriously for 'the product of their labour is by them consider'd of more importance, and the interest of the child is sacrificed to the selfishness of the parent'. William Thomas noted that the mining communities were generally healthy, although in old age the miners did suffer from asthmatic complains. He added that medical care is sometimes provided by the colliery owner and otherwise by the weekly contributions of the miners. Friendly societies existed to provide for the families in sickness and old age. He commented that men, who had availed themselves of the opportunities to learn reading, writing and arithmetic, could rise through the ranks to the position of colliery viewer which was a stepping stone to colliery ownership itself.

Mining was an occupation marked by dangerous working conditions and high earnings. These were spent in high living in which gambling and drinking played a prominent role. Edward Chicken's poem 'The Collier's Wedding', which culminated in the guests being 'dead drunk', epitomises the spirit of the collier lads who 'got money fast' and 'had merry days while it did last; did feast, and drink, and game, and play, and swore when they had nought to say'.[116] Excessive eating, drinking and gambling, disregard for education, and ungodliness, were features of the mining communities well into the

---

[114] Huntington Library: MO 5840, Elizabeth Montagu to Sarah Scott, 17th July 1766.

[115] NEIMME: Tracts 29 p. 234 – 263 Letter from William Thomas to Sir John Swinburne.

[116] Edward Chicken (1698 – 1742) was parish clerk at St. John's church in Newcastle. His poem tells the story of a Benwell pit lad's courtship and marriage. NEIMME: Bell/11/601

nineteenth century as the early histories of Methodism in the region testify. But that was only part of the story.

Author's collection
Fig. 18: A Miner and his Lass near Wylam Colliery

There were many sensitive and intelligent men from the coalmining communities who rose to positions of prominence and a few to great eminence. The distinguished mathematician Dr. Charles Hutton was born in Newcastle and worked with his brothers at Longbenton Colliery, where his father was deputy overman; George Stephenson, who worked as an enginewright at Killingworth, became one of most distinguished railway engineer of his age; arguably the greatest mining engineer of the nineteenth century, John Buddle Junior, was introduced to work underground at the age of six by his father; Geordie Elliot began work as a trapper boy at Penshaw Colliery and became President of the Mining Institute, M.P. for North Durham and a friend of Prime Minister Disraeli and the Khedive of Egypt. In the early nineteenth century, a Percy Main miner, Joseph Skipsey, began his working life at the age of seven as a trapper boy at the local colliery. His literary success gained him the epithet of the pitman's poet and recognition from distinguished writers of the calibre of Oscar Wilde. Many other worthy men did not achieve recognition at a national level but were respected leaders of the Methodist, Co-operative and Trade Union movements within their local communities. Clearly, men were fashioned in very different ways by their experience down the pit, what was often referred to in the coalfield as 'the hard school of life'.

Although the men were free to leave at the end of the year, they were liable to legal action if they broke the terms of their bond and left beforehand. Notices were placed in the press and when the whereabouts of the culprit was ascertained more direct action was taken as Brown's letter to Mr Errington indicates.

'I understand John Ursop one of my bound servants is employed by you and now working in some of your Collieries contrary to his agreement with me and the laws of this country – you'll be so obliging as discharge the man and prevent my commencing an action against him in order to compel him to do what an honest man ought to do without compulsion'.[117]

William Brown also encountered problems with men absconding while he was managing Walbottle Colliery as is illustrated in his letter to the Duke of Northumberland dated 16th February 1771.

---

[117] NEIMME; WBLB II p. 2 William Brown to John Errington, 5th January 1765

> 'We had a Fire on the 30th Ulto by which one Man lost his Life and two more were slightly burnt, that Fire (with some we had before) has given some of our Men a preference (at least) for Fear, and some have run off and left Us, the Loss of which together with the Loss of some that Listed on Board the Tender and others that had left us a few days after they were Bound, has reduced our Quality of Hands very much, I have consulted Mr Fawcett how to bring these loose Fellows to a sense of their Duty and am proceeding against them accordingly – a want of Authority, and Subordination at the Colliery has been a great loss.'[118]

The lengths to which William Brown was prepared to go to retrieve miners who had broken their bond to Walbottle Colliery is vividly described in William Gibson's view book. No doubt Brown was prompted by the Duke's criticism of his management of the colliery and in particular his inability to deliver top quality coal at an acceptable price.[119] On Monday 11th March 1771, his assistant was sent out in search of the deserters. For a week in the bitter cold of late winter, Gibson devoted all his time and considerable energy in hunting the absconders. First, he travelled to Chester le Street to get a warrant from Captain Milbanke and then on to various pits in County Durham before ending the day at West Auckland at 10 p.m. On Tuesday, he searched at pits in the Bishop Auckland – Cockfield Fell area before reaching his lodgings at Cornsay at 5 p.m. This forty mile journey was a 'disagreeable ride, the snow being so deep upon the ground'. On Wednesday, he travelled from Cornsay via Shotley, Hexham and Haltwhistle to Glenwelt. The next day he journeyed via Carlisle to Wigton and on to Cockermouth. On Friday, he found two of the men at Banklands Colliery near Workington, before proceeding to Whitehaven, where the following day he found another four of his men. On Sunday, he set out from Wigton for Newcastle where he arrived at 10 p.m.

However, by 1771, the Duke was not happy with William Brown's management of Walbottle Colliery. The cost of the development was £12,000, twice the original estimate provided by William Brown. Also, the Duke was concerned that the pay bills indicated that the production costs were 9s per chaldron, which needed to be brought down to 6s otherwise he would suffer an estimated annual loss of £2,000. Furthermore, Brown had enemies at court: someone drew up a document for the Duke, using extracts from Brown's letters, which concluded with the acerbic and damaging remark – 'It will be apparent how inconsistent and contradictory Mr. Brown's Assurances have perpetually been and how little they could be relied upon.'[120]

The outcome was that Brown lost the contract for managing Walbottle Colliery which was assigned to William Cramlington from September 1771 for 21 years. His contract provides some very valuable information about the operation of the colliery, including the minimum and maximum output, and the fact that it was intended to mix the better coal from the Engine seam with the inferior coal from the Main seam as a means of increasing the colliery's sales to the London market.

---

[118] The archive of the Duke of Northumberland at Alnwick Castle K/I/17/i/9/3 William Brown to Duke of Northumberland, 16th January 1771. By 'fire' Brown meant an explosion of methane gas.

[119] The archives of the Duke of Northumberland at Alnwick Castle K/I/17/i/1/1

[120] The archives of the Duke of Northumberland at Alnwick Castle K/17/I/9/11

'The said Mr Cramlington contracts and engages to work and lead out of the said Colliery not less than 800 Tens of Coal each Ten to contain 17 chaldrons and a half and each chaldron 24 Bolls Newcastle measure and no more than 1,000 such Tens without license from the Duke annually at the rate of five pounds fifteen shillings per Ten for all such Coals as are fit for the London Market or to be vended Coastwise, the same to be as good as the nature of the Mine will admit of, and sufficient quantity of the Coal of the Main seam now in working being always properly mixed with the upper seam so as to make the whole of good quality and most fit for the Market, and at the rate of five pounds five shillings per Ten for all inferior and small Coals that shall be so worked and led. The Contractor agrees that the above mentioned allowance or rates…shall be accepted by him in full satisfaction and discharge of all expenses for working, drawing, loading and delivering the said Coals and that he will not draw to Bank or bring to account any refuse Coal that shall be objected to by the Dukes Agent but for landsale only'. [121]

Also within the contract was the stipulation that 'the Duke and Duchess' Tenants shall be obliged to lead the Coals at a reasonable price not exceeding eight pence per waggon of 24 Bolls each', slightly more than a chaldron measure. The overweight was to allow for spillage since the waggonmen were required to deliver a full Newcastle chaldron (53cwt) at the staith. The tenants' contracts stipulated that:

'The Duke and Duchess reserves a power at any time during this present demise to oblige the Tenant to serve them with Horses for such Waggons as shall be thought necessary…at the usual rate per Waggon in proportion to the length of the Waggon Way belonging to Walbottle Moor Colliery'.

Clearly, the tenants on the estate were expected to provide the horses and waggonmen for the railway. However, it is important to recognise that although service on the waggonway was an obligation of the tenants, it was also a lucrative source of income and no doubt welcomed.[122] The operation of the Throckley and Walbottle Waggonway is considered in more detail in the next chapter.

---

[121] The Archives of the Duke of Northumberland at Alnwick Castle K/17/I/1/2. The output was to be between 37,000 and 46,000 tons.

[122] The Archives of the Duke of Northumberland at Alnwick Castle K/17/I/1/9/5, 1/9/6 and 1/9/9.

Number of Waggon Way Horses to be kept by the Tenants of Newburn and Walbottle

Name	Farm	Horses	Name	Horses
Ralph Davison	Chaple House	3	Heirs of the late Thomas Pearson	1
Wm Longridge	Graham Farm	2	Executors of the late William Bell	2
Henry Forster		4	John Soulsby	0
Geo. Dobson	Pigs Hall	2	John Rutter	½
Wm Longridge	own farm	2	Robert Riddel	½
Thomas Green	Hill Head	2	Thomas Bell and John Lennox	¾
William Austin		½	Jn. Birtley	¼
Thomas Softley		½	Cramlington (Walbottle)	2
Cramlington (Newburn)		8	Total	31

Chapter Six: The Throckley and Walbottle Way – William Brown's Railway

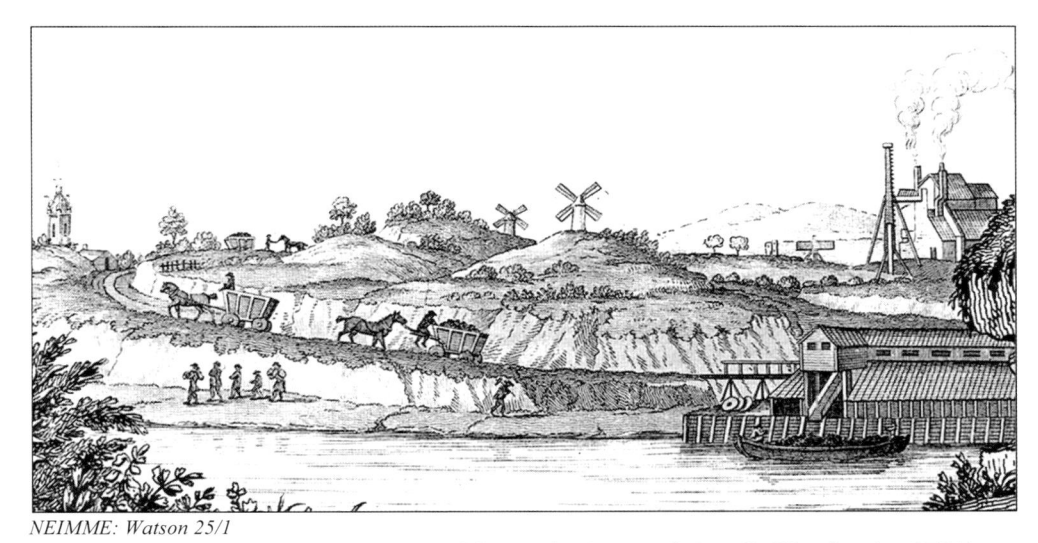

NEIMME: Watson 25/1

Fig. 19: Cartouch from John Gibson's Plan of the Collieries in 1787

William Brown's former apprentice, John Gibson, captured the industrial scene on Tyneside in the cartouch for his 'Plan of the Collieries on the Rivers Tyne and Wear': he shows the waggonway providing the vital link between the colliery and the staith. A keelboat is preparing to ferry its cargo downriver to the colliers moored at Shields from where it would be shipped mainly to London and the South East. Waggonways were an essential part of the transport system for getting the coal to market and without them the fourfold expansion of the coal trade in Northumberland and Durham during the eighteenth century would not have been possible. Gibson's cartouch illustrates how this wooden railway functioned: after his journey from the colliery along a reasonably level route, one waggonman has uncoupled his horse at the top of the incline in preparation for the steep descent down to the staith; another, with his horse attached to the back of the chaldron waggon to keep it out of harm's way, is undertaking the dangerous task of controlling the loaded waggon down the steep hillside; while a third waggonman returns back to the colliery with an empty waggon. Such were the railways of Brown's time.

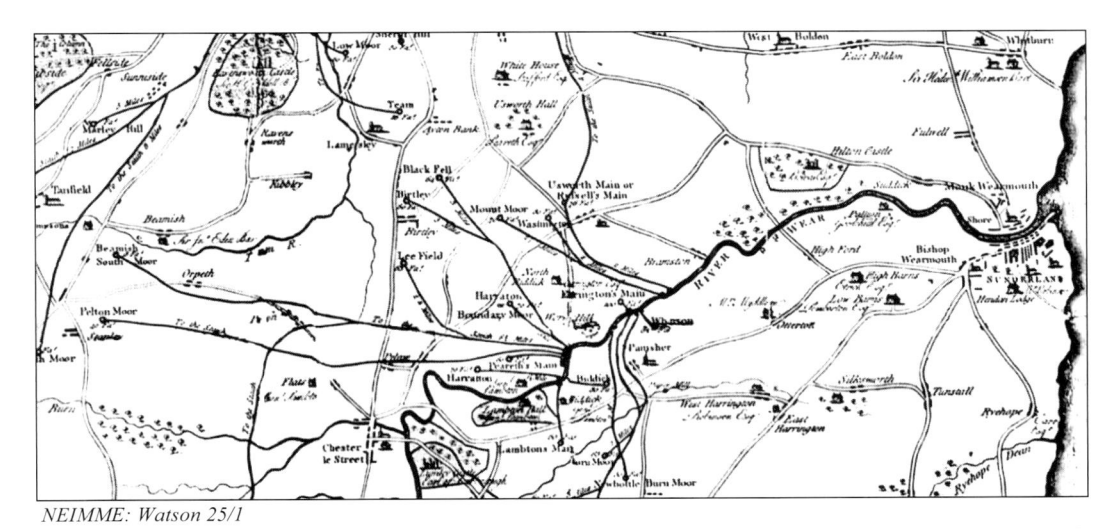

NEIMME: Watson 25/1

Fig. 20: Part of John Gibson's Plan showing Waggonways on Wearside in 1787

Gibson's map shows a complex network of waggonways in the region. Not all are well documented but fortunately there is sufficient evidence surviving for Brown's Throckley and Walbottle Way to provide an understanding of what railways were like in the mid eighteenth century. The Throckley Way of 1751 is unlikely to have been William Brown's first waggonway for by 1754 his reputation as a builder of railways was sufficient to secure a contract to build a line underground at Bo' ness Colliery in Scotland. Later he was responsible for building most of the lines in the Tyne Basin (which are described in the chapter seven) and he also built several lines to the River Wear including the Beamish Way, a rival to the famous Tanfield Way.[123]

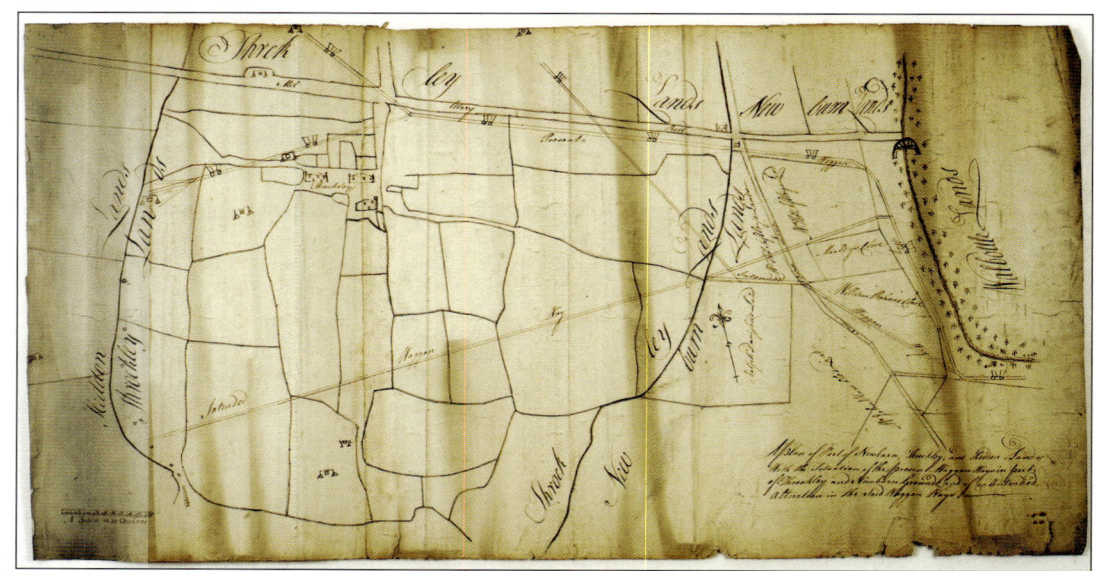

Courtesy of the Archives of the Duke of Northumberland at Alnwick Castle, AC:O.XVII.4.2.
Fig. 21: Throckley Waggonway in 1755

Waggonways were expensive and in the mid eighteenth century landsale collieries were not able to afford the large outlay of capital needed for their construction. The decision to make Throckley a seasale colliery justified the building of its first railway. Opened in 1751, the line ran from the pits near old Throckley village, down the south side of the turnpike road to Walbottle Dene, where it turned south and adopted a steep descent crossing Walbottle Dene on a viaduct en route via Newburn village to Lemington (fig.21). This map was drawn, probably by Brown himself, to seek permission from the Duke of Northumberland for a re-alignment of the line in Newburn estate brought about by the opening of Heddon Colliery in the west and the new pits for Throckley Colliery to the north. Brown also sought to adopt a more gentle descent to the bridge. The main engineering feature on the line was Throckley Bridge, a huge wooden structure 84 yards in length, spanning Walbottle Dene on twenty three pairs of geers or wooden trestles.

The Wylam Way was opened in 1756. It joined the Throckley Way east of Newburn Hall which it used through Newburn before terminating at a staith on Mrs Montagu's land further east in Lemington. When the first fire engine was erected in the south west corner of the Throckley estate in 1760, a branch was laid from the pits in south Throckley to the Wylam Way, which the south Throckley Way joined near the wash hole at Throckley Reigh.[124] After the erection of the second and third fire engines in the

---

[123] Les Turnbull 'Railways Before George Stephenson' p.67-73

[124] NEIMME: Watson 2/12/10. The subject of washholes is discussed in detail later.

south east corner of Throckley, production was concentrated in the southern part of the estate. Mining for seasale coal may well have ended temporarily in the northern part, resulting in the closure of the original waggonway from the north. Certainly, by 1766, the bridge across Walbottle Dene was in need of repair and the decision to build a fourth engine to win the Splint coal seam north of the main fault, made a full renovation necessary. Christopher Bedlington recorded on Monday 7th February 1767 that 'the Bridge across Walbottle Dene is now so far finished as the Waggons can come down to the Staith'.[125] It is likely that this was also the occasion for abandoning the section of line used jointly with the Wylam Way and replacing it by a new section of line further north for the exclusive use of Throckley Colliery.

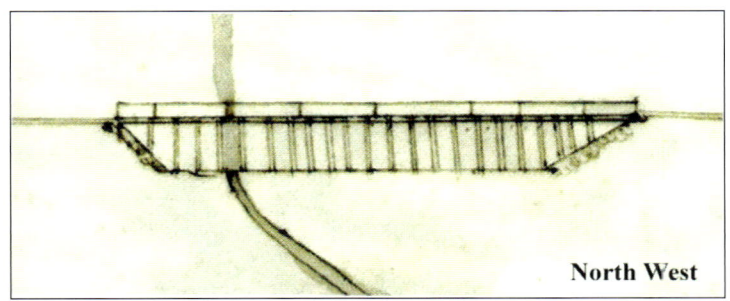

Courtesy of the Society of Antiquaries of Newcastle upon Tyne
Fig. 22: The Bridge over Walbottle Dene

The opening of Walbottle Colliery in 1770 was to bring about other major changes to the railway. A fourth staith was built at Lemington for Walbottle coals immediately north east of Throckley staith and the Throckley Way was reorganised to accommodate this new colliery. A new main line through Newburn, known as the Partnership Way, was built for both and the collieries and the Duke purchased a half share in the section of the Throckley Way from Walbottle to the river which was henceforth used by both collieries as a bye way for returning waggons. New branches were built from both the Engine Pit in Newburn and the Betty Pit in Walbottle. At the same time the Wylam Way was altered by straightening the line of the route through Newburn.

An estimate for 'the alteration to Throckley Waggon Way in Newbourn Hall Estate' reveals the cost of laying a yard of track in 1769:

Preparing and levelling by the yard at 4d	4d
2 sleepers at 10d	1s 8d
2 yards of Fir Rail and 2 yards of Beach	1s 8d
To laying	2d
To Pins	2d
To Ballasting	3d
Total	4s 7d per yard

The total cost of the laying the new waggonway comprising 1,900 yards of main way and 1,000 yards of byeway was estimated at the conveniently round sum of £500, which had been kept low by recycling 100 yards of main way and 600 yards of byeway. The costs were to be shared between the Duke, owner of Walbottle, and Bell and Brown, owners of Throckley. An interesting if somewhat dubious scheme was put forward to fund Bell and Brown's share at the expense of John Blackett, the owner of the Wylam Way, and the waggonmen. The size of the Throckley waggons was to be enlarged. By

---
[125] NEIMME: East 10B/7.

increasing the capacity of the waggons by 11%, from 20 Bolls (193,576 cu.ins.) to a full chaldron (217,989 cu.ins.), and retaining the leading price at 10d per trip, it was estimated that a saving of £500 would be made over a five year period on the 12,000 waggon journeys along the route each year at the expense of the Duke's tenants. Furthermore, since fewer waggons were used to move the same amount of coal, there was an additional saving of £291 – 13 – 4d on the cost of wayleave rent to Blackett for the use of 1,000 yards of the Wylam Way.[126]

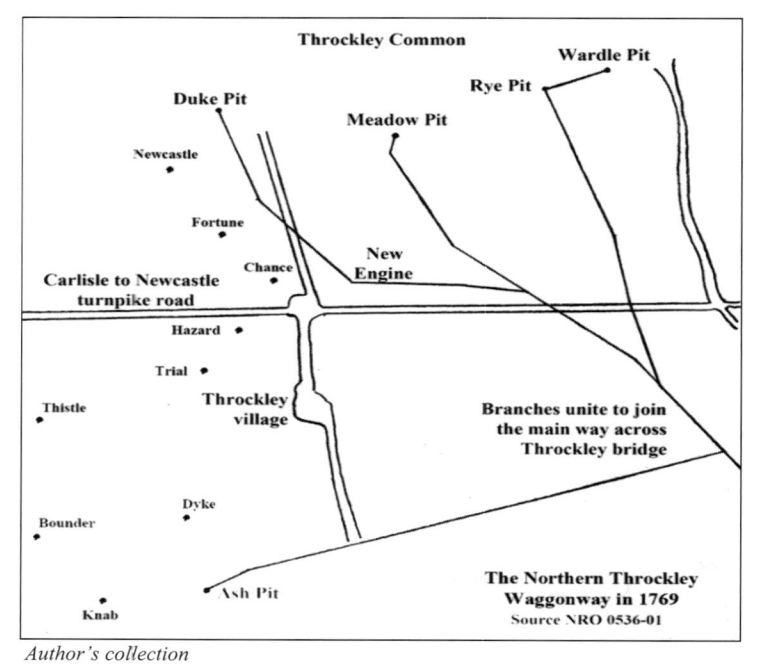

Author's collection

Fig. 23: Throckley Waggonway in 1769

Figure 23 shows that the changes requested in 1755 had been put into effect. However, by 1769, the link with Heddon Colliery had been severed, although part of the line was still in use to serve the Ash Pit of Throckley Colliery. It is likely that Heddon Colliery was gaining access to the South Throckley Way by a tortuous and dangerous route down Heddon Banks. The map, when compared with figure 21, also illustrates an important feature of railways at this time when collieries consisted of numerous pits which were worked for about six years. The branches of the railway changed as one pit fell into disuse and was replaced by a new one. The area of mining was like a large marshalling yard with branch lines leading to the main line to the staith. When collieries became deeper there were fewer pits with more permanent rail connections but then railways were built underground to marshal the coal. Because of the number of pits, it is very likely that every field in Throckley estate had a railway line running through it at sometime during the operation of the mine in the eighteenth century!

Waggonways were constructed and maintained by specialist contractors such as the Forsters of Newburn Hall who were doubtlessly Brown's contractors for the Throckley line. No maintenance contract survives for the Throckley Way but Richard Forster's proposals for the maintenance of the Brown family's other waggonways at Bigges Main and Willington still exist. The tender for April 1788 is quoted in full as an example of the practice on an eighteenth century waggonway:

---

[126] NEIMME: Watson 2/10/3

> '1st We propose taking the above Waggonways & Waggons to keep in repair by the Ten of 22 Waggons to the Ten at 2s 8d per Ten from the Pits now working or from any other Pit or Pits in proportion according to the length of the way from which the same may be led.'

Thus Richard Forster was paid according to the production of the colliery and, on an annual vend of 30,000 chaldrons (1,590 tenns) at Bigges Main, his income would be £212. If other pits were brought into production an additional rate would be determined by the distance the waggons had to travel from the individual pits.

> '2nd The undertakers to find Workmanship and all Materials. Smith work etc. included in the above price.'

> '3rd The undertakers to be paid for all back Carriage etc. according to the length of the way.'

The was no payment either to the drivers or the maintenance staff for empty waggons returning but when the waggons carried a cargo on the return journey – such as timber and bricks for the colliery or lime and manure for the farms – additional payments were made. On the return journey the waggons usually carried a half load since they were generally operating against the gradient for most of the journey.

> '4th If any part of the Way or Waggons be unoccupied for Six Months to be repaired at the Owners Expense.'

In the event of the waggonway being closed, for example as a result of an accident which disabled the colliery, the undertakers would not be responsible for the repair since they had thereby lost their source of revenue, the tentale payment, or the money paid for the number of tens carried by the waggonway.

> '5th The undertakers to keep in repair Cutts, Batteries and Waggon Way Gates in the said Way.'

Cutts and batteries were in today's parlance cuttings and embankments. The waggonway frequently divided the lands of neighbouring farms and gates were placed at appropriate points to provide official crossing points. Elsewhere, the railway was enclosed on both sides by fencing, a thorn hedge and a deep ditch to prevent trespass, especially by stray animals. The ditch also provided drainage for the railway. Many of these ditches and hawthorn hedges survive today.

> '6th The Undertakers to be found with house and fireing by the Owners.'

> '7th The Owners to find the Undertakers a Horse or Horses to Ballast the Waggon Way etc.'

> '8th The Undertakers will engage for the Term of Seven Years if agreeable to the Owners.'

'9th Providing the Owners be agreeable to let the two places both together, We propose taking Willington at 3s 3d per Ten and Benton at 2s 7d per Ten.'

Richard Forster and William Eltringham were given the maintenance contracts for the waggonways and waggons at both Bigges Main and Willington collieries. The probability is that they also maintained Throckley and Walbottle. Later tenders include interesting items like 'ashing the runs', particularly necessary in bad weather to provide adhesion for the waggons descending a hill, and 'assist in shovelling snow' a reminder of the severity of the winter in the eighteenth century when even the River Tyne was known to freeze over. Indeed, on more than one occasion ice damaged Lemington staiths. Men were employed as creasers to keep the rails clear for the free passage of the flange or crease as it was known at the time.[127]

The autobiography of Anthony Errington (1778-1825), who had the contract for maintaining the Brandlings' waggonways at Felling and Gosforth collieries, provides some interesting detail of the working conditions of a waggonwaywright in the later part of the eighteenth century. Like his father, Anthony was responsible for both building and maintaining the waggonways and the waggons for which he was paid about £1 per week – slightly more than the hewers at the colliery. Besides the railway on the surface, they were responsible for the railways underground, which entailed working two or three nights a week on the maintenance shift at the pit. Their work was not without its dangers both above and below ground. Anthony Errington records that his father 'had his Coller Bone Broke in the Ann pit by a waggon running amain' and his ribs were broken six different times. On a more positive note, he also claimed that his father was 'the first to putt 2 Convoys upon the pit waggons' to improve braking and 'the first that invented the Double Switch' for the track. However, it is difficult to substantiate these claims.[128]

In December 1771, John Marley made a survey of the Walbottle Way for the Duke and Duchess of Northumberland probably on the occasion of Brown being replaced by Cramlington. This survey, together with the schedule accompanying the map and the colliery accounts, provide what is probably the most comprehensive description of an eighteenth century waggonway to survive.[129] The documents reveal that there were four different categories of railway. The most important was the double main way for the loaded waggons which was built of two rails, one on top of the other. Throughout the waggonway the upper line was made of beech, a wood which was available in long straight lengths. Beech was a close grained wood which was reputed to wear smooth, thereby decreasing the friction of the waggon making heavy loads easier to haul. The lower rail was mainly oak (72%) but in parts of the line fir, a cheaper alternative, was used (28%). A drawing of a main way exists in John Buddle's note book and an example was excavated at Carville in 2013 about 100 yards from his house (fig.24). The return route for the empty waggons, the bye way, was always single rail made of fir. Less attention was paid to levelling this line since it usually carried only empty waggons: as Brown explained in a letter to his partner Masterman there was 'no necessity of having

---

[127] NEIMME: Watson 2/214; 3/4/12, 13, 20.

[128] P.E. Hair editor 'Coals on Rails' Liverpool University 1988. The Blackett papers record payments to Elizabeth Scott for making and reparing waggons for the Wylam Way – see A. Clothier 'Beyond Blaydon Races' p.63.

[129] Northumberland Record Office: Sant/Beq/09/01/01/31. NEIMME: Watson 2/7/57-65,139 – 160. Alnwick Castle K/I/17/i/8/2.

a Byway to run so much upon a Horizontal plain as the Main way'.[130] The same letter states that 'a Main way is about a yard or four feet Distance' from the bye way, which is different from the practice on the later iron railways, where the space between two lines was a minimum of two yards. This shorter space was possible because the chaldron waggon had wheels on the outside of the body. No mention is made of the gauge of the Throckley line which was probably five feet.[131] The branches to the Betty and Engine pits, which carried both loaded and empty waggons, were described as 'main single oak'. This probably meant that the sleepers were laid at two foot intervals, as for the main way, but only a single rail was laid, as on the bye way. The fourth type of rail was the false rail, or guide rail, which was used extensively throughout the waggonway. It is marked as dotted lines beside the track on the map of the line (fig.25).

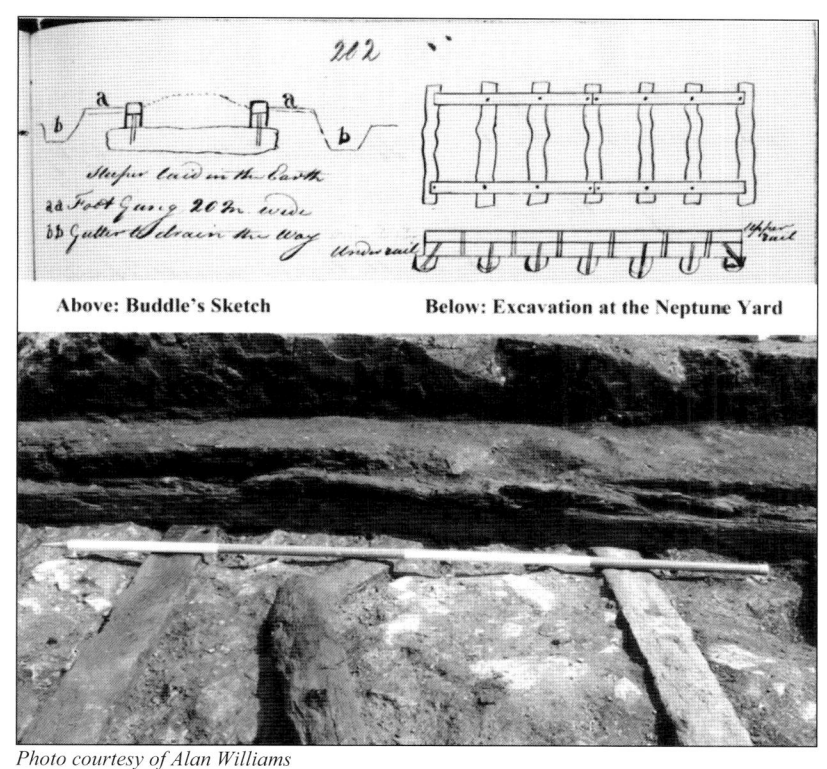

Fig. 24: The Main Way at Carville

The accounts record the delivery of large quantities of imported timber: John Snowden was paid 'for freight of oak timber and pumps from London to Shields: £9 – 6 – 6d'. Another William Brown and his colleague Peter Findley, two skippers of keelboats, were paid £1 – 6 – 8d for ferrying this cargo the fifteen miles upriver from Shields to Lemington. The timber had to be cut to size and there are many references like the one recording 'Thos Oliver for sawing 3½ tons of Waggon Way Rails at 3s 9d per ton'. Rails also arrived ready made: Thomas Story delivered a consignment of rails and sleepers in the spring of 1769. Robert Forster was the man principally responsible for laying the track: he was paid 2½d for laying ordinary rails and 2d for false rails. Nicholas Chancer, a mason, was responsible 'for making the Cutts and Batteries of the New Waggon Way from Lemington through Newbourn Hall grounds' and he also supervised the building of the new staith at Lemington.

---

[130] NEIMME: WBLB I p.302, William Brown to Henry Masterman, 30th November 1754.
[131] There was no standard gauge for the waggonways since they operated in clusters independent of eachother: the Tanfield line was 4 ft.; the Beamish Way 4 ft. 4 in.; the Willington Way 4 ft. 8 in.; Heaton Main Way 4 ft 10 in and the Wylam line 5 ft.

55 - switch

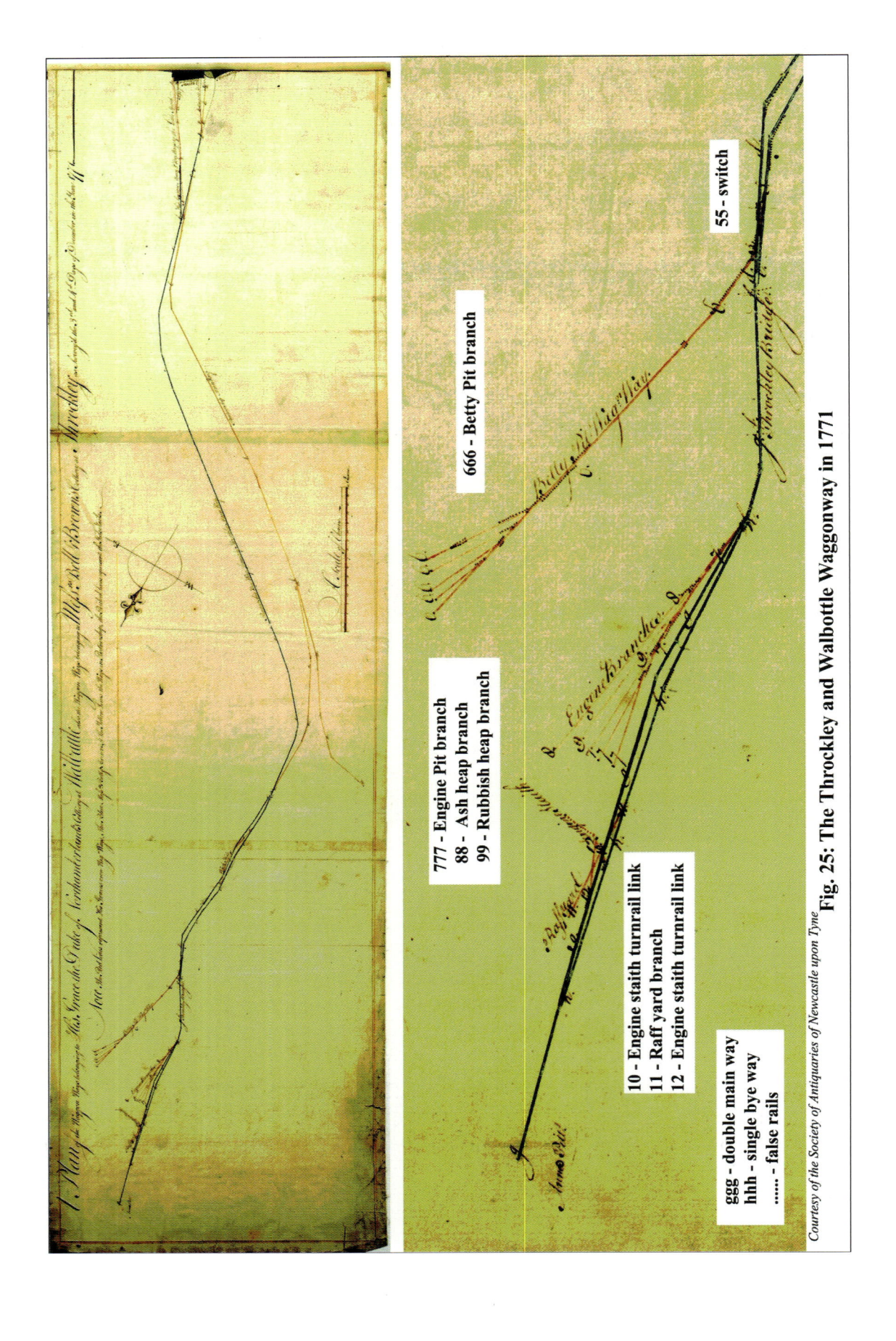

666 - Betty Pit branch

777 - Engine Pit branch
88 - Ash heap branch
99 - Rubbish heap branch

10 - Engine staith turnrail link
11 - Raff yard branch
12 - Engine staith turnrail link

ggg - double main way
hhh - single bye way
...... - false rails

Courtesy of the Society of Antiquaries of Newcastle upon Tyne **Fig. 25: The Throckley and Walbottle Waggonway in 1771**

In the schedule, there is no mention of sleepers: these would have been invisible when the survey was made since they were buried by the ballast laid over the sleepers to prevent wear from constant trampling by horses. However, Brown's letters contain a reference to the purchase of ships timbers for sleepers. In February 1765, William Brown wrote to Captain Snowden 'I'll take it kind if you'll call at Mr Cox's and Mr Trinders Shipbreakers at Cuckolds Point and find if they have any sleepers'.[132] Re-used ship's timbers were found at the excavation of the Bigges Main Waggonway. Oak was generally used for the sleepers and there is a reference in the colliery accounts of a payment to 'Thos. Hudson for 244 oak sleepers at 9d each' but whether these were new or re-used timber is not known. However, Anthony Harrison supplied '100 Ash sleepers at 6d' as a cheaper alternative for the branch to Little Spout. Some sleepers were of a very rough construction as figure 24 shows.

The western section of Brown's plan of the Throckley and Walbottle Way (Fig.25) depicts a complex layout of tracks. The double main way (ggg) and the single bye way (hhh) of Bell and Brown's original waggonway from Throckley are shown running from the site of the newly sunk Ann Pit of Walbottle Colliery, past the Engine Pit, to the bridge across Walbottle Dene. The main way was built of oak with beech rails on top; the bye way was single fir rails. A branch line (7777) led from the west end of the bridge to the Engine Pit heapstead where there were three sidings. Two branches ran from this line: one to the ash heap (88) and the second to the rubbish heap (99). Another two branches led from Throckley bye way: one to the engine staith turnrail (10) and another to the raff yard (11). Finally, a branch is recorded from the Throckley main way to the engine staith turnrail (12). Dotted lines indicate the position of false rails.

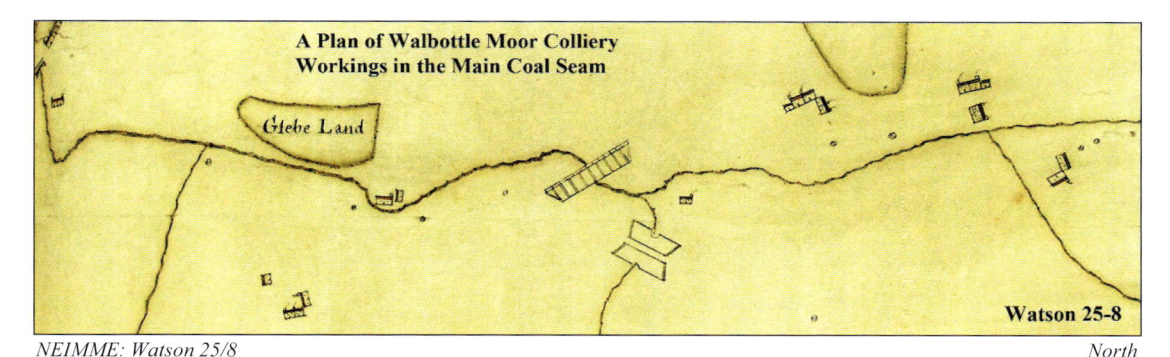

NEIMME: Watson 25/8   North

Fig. 26: The Two Bridges at Walbottle

The colliery was divided by a deep ravine: the Engine Pit was to the west in Newburn and the Betty Pit to the east in Walbottle. A large wooden bridge carried only the double main way across the dene. The main way was built with fir bottom rails and beech top rails. An earth embankment with a culvert through for the stream carried the branch line to the Betty Pit over a tributary stream. These were the two common types of bridges on the waggonways: stone bridges like the Tanfield Arch were the exception.[133] This second bridge was the scene of a major disaster on 24th July 1787 when the culvert was blocked following a storm and flash flood. The embankment acted as a dam which eventually burst and, in the words of the local historian John Sykes, 'with an impetuosity scarcely conceivable, instantly carried away an adjoining mill' and 'three

---

[132] NEIMME: WBLB II p. 19, William Brown to Captain Snowden, 11th February 1765.
[133] A small stone bridge which once carried the Wylam Way over the New Burn still survives.

houses at the east end of the village of Newburn' where 'all the houses in the low part of the village where filled with water'. Four people lost their lives.[134]

On the eastern side of the main bridge, a short link line led from the Throckley bye way to the Throckley main way, where the only switch on the entire network was recorded (5 5). Although there were many junctions in the system, there is only one mention of a switch or point. This would imply that in general the drivers were expected to manhandle their waggon over a junction. The purpose of the switch was probably to admit traffic from the Betty Pit branch onto the main way. Iron plates, which were expensive items at this time, protected the switch rail from wear. A branch line led from the switch to the Betty Pit heapstead where there were four sidings (6666). Later, this branch was extended northwards and became the main outlet for the collieries to the north as is shown on Gibson's plan of 1787 (fig. 1). By that date, William Brown was dead and his son had abandoned Throckley to concentrate on the larger collieries at Willington and Bigges Main collieries in the Tyne Basin. Joseph Lamb and Partners took over the lease of Throckley but they mined further north on Greenwich Moor and the branch across the bridge to Brown's Throckley Colliery was closed. By 1787, the waggonway from Holiwell had also been altered to join the Walbottle Way.

Courtesy of Alan Clothier

Fig. 27: False Rails at Lambton 'D' Pit

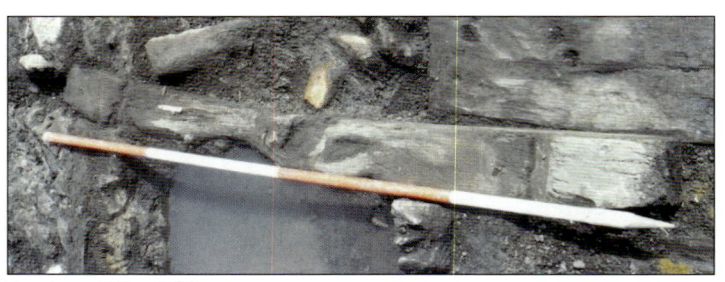

Courtesy of Alan Williams

Fig. 28: Re-used Ships Timbers

---

[134] Sykes, J., Local Record Vol. I p.344.

The valuation of 1770 records the rails and sleepers held in reserve. Like the wooden waggon wheels, the wooden rails were subject to much wear and tear making it necessary to keep a significant quantity in stock for the maintenance crew.

Laying near the waggonway and Branches		
108 yds of Beach Rails	at 5½d	£2 – 09 – 06
50 yds of Oak Rail	at 8d	£1 – 13 – 04
197 oak sleepers	at 8d	£6 – 11 – 04
20 yds of Fir Rails	at 5d	£0 – 08 – 04
33 ash sleepers	at 5d	£0 – 13 – 09
420 yds Beach Rails at Staith	at 5d	£8 – 15 – 05
60 oak sleepers	at 8d	£2 – 00 – 00

By the beginning of March 1770, coal was being mined at the Engine Pit and the Duke's waggons were forced to use the Throckley Way to transport it to Lemington since the New Waggon Way, the Partnership Way, was not finished until 24th September 1770. The Duke had to pay Bell and Brown for this privilege. He also had to pay £70 'for leading materials to the Colly at Walbottle viz Timber, Iron, Brick, Lime and damage by the great weight of many of the materials'.[135] This serves as a reminder of the important part played by the waggonway in the opening up of the colliery which is the reason why the railway was often in operation before coal production began. The valuation also states the types of waggons in use at that time and their value:

7 Coal waggons with 2 iron wheels	£10 – 10s	£73 – 10s
1 coal waggon with 4 wood wheels		£8 – 08s
2 pairs of new iron wheels	£4 – 10s	£9
2 ballast waggons with wood wheels		£6
1 coop waggon with wood wheels		£2 – 10s
		£99 – 08s

It is interesting that nearly all the coal waggons were the new type with two iron wheels at the front and wooden ones at the back. The eight coal waggons in the valuation represented only about half the number that would be required when the colliery came into full production. These waggons held a Newcastle chaldron (53cwt) but this was not always the case. Thomas Donnison's colliery at Birtley, for example, used 46 cwt waggons on his waggonway to the Tyne and 42 cwt waggons to lead to the Wear. Furthermore, the capacity of the waggon could be altered to suit the needs of the market. Buddle's diary for the 26th May 1816 records: 'As Heaton measure is proved to be worse than any of the other collieries, fixed with Mr Potts to put ledges upon the waggons by nailing an Inch Deal upon the fore and side overings – tapering them off at the hind end. By this small addition it is expected that the measure will give content'. The number of ballast waggons indicates that ballasting was a regular activity made necessary because the ballast was disturbed by the continuous tramping of the horses. Likewise, the job of removing the ballast from the edge of the rail (known as creasing) to accommodate the flange of the wheel (the crease) was an on-going task. It was also expensive. The coop waggon was probably used to supply materials to the colliery.

---

[135] NEIMME: Watson 2/7/69.

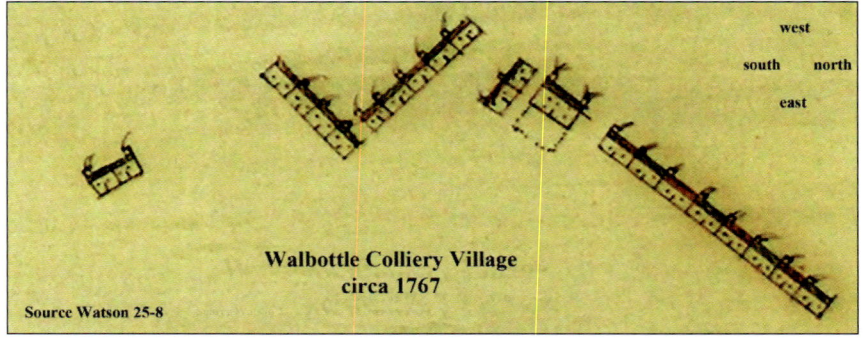

NEIMME: Watson 25/8

Fig. 29: Walbottle Colliery Village circa 1767

The colliery houses at Walbottle were situated to the west of the old village near the Betty Pit and comprised of a large two story building, which was probably occupied by the manager; and 21 single story cottages for the miners. The floors were probably flagstones, there would be an outside dry toilet and water would be supplied from a communal pump. A later colliery row (marked on the First Edition O.S. map as New Row to the north of the cottages in Watson's map) consisted of back to back houses comprising a living room and pantry downstairs, and a bedroom in the attic, which was accessed from a ladder in the living room; and it is likely that the houses shown above were similar. No record survives which record the size of the population of the village in 1767 but a survey of the mining accommodation in Denton in 1808 revealed an occupation level of seven persons per household. However inadequate this accommodation may seem to the modern reader, it was substantial for the time and decent housing was one of the incentives which attracted a labour force to the area.

Courtesy of the Society of Antiquaries of Newcastle upon Tyne: NRO: SANT-BEQ-09-01-01-31

Fig. 30: The Wash Hole at Newburn

The next section of the Throckley Way was a descent from the bridge across Walbottle Dene to Newburn. The main way, 509 yards long, had oak bottom and beech top rails (dddd); and there was 106 yards of false rail reflecting the difficulty of the route. The single bye way, 475 yards long, had fir rails (eeee); and there was 137 yards of false rail. A branch to the wash hole, 158 yards long, was recorded. The location of the wash hole appears curious (the branch ran from the bye way, across the main way and returned to the bye way) suggesting that the wash hole was an earlier feature. The pond, which was more necessary for the Throckley waggons following their steep descent from the moor than the Walbottle waggons which had just started their journey, was probably part of the original layout of the line when there was only a single main line down the east side of the dene. The latter addition of a new main way and the conversion of the original main way to a bye way to accommodate the traffic from Walbottle Colliery created this curiosity. The purpose of the wash hole was to wet the wooden wheels of the chaldron waggon to prevent them drying out and cracking which would usually be undertaken on the return journey. William Brown was consultant to Newark

Colliery in Fifeshire and the drivers on the waggonway to Pittenweem harbour were required to run their empty waggons through the wash hole on each return trip to prolong the life of the wooden wheels. Perhaps this was also expected of the men at Walbottle and Throckley but it was not mentioned in their contract. The gradient on the upper part of the Throckley line was a steep one – on average 1:28 – which would test the skills of the waggonmen. The friction of the wooden brake on the wooden wheel would cause heat and possibly a fire – hence the need for the wash hole. [136]

Courtesy of Alan Williams

Fig. 31: Wash hole on the Willington Waggonway

The excavations at the Carville in 2013 revealed details of the construction of wash holes for the first time. A loop, about thirty metres in length, branched off the main way and ran downhill to the pond before returning uphill to the main line. There was a raised central walkway for the horses. The area also had a loading platform for return cargoes

---

[136] NEIMME: Watson 2/12/20

and some curious trenches which may have been watering troughs for the horses: the facility was more than a pond and is perhaps better described as a service station. It is interesting that the field in which it was situated was known as Well Laws: the water supply was probably a spring or a small stream marking the boundary of the estate.

The gradient of the waggonways was the subject of much discussion between Carlisle Spedding and William Brown for controlling a loaded waggon, weighing about four tons, on a falling gradient such as the descent from the Military Road to Newburn, was a major problem. Brown explained the practice in the North East and provided some interesting detail about the railways of the mid eighteenth century.

> 'As to the Runns in our Waggon ways, I know of none that Exceeds 3 inches in a yard and very few that is so much for we find that our Convoys will not hold a Large Waggon when the Runn is so much Notwithstanding we lay ashes and sometimes Cynders to Ruffen the Railes'.

Brown's solution to steep descents (greater than 1:12) was to use smaller waggons. He added that 'we always have wood on the hind wheels' and that by means of a long convoy the drivers were able to trail all four wheels.[137]

Author's collection
Fig. 32: The Chaldron Waggon

In a letter to the London Magazine, dated December 21st 1763, a correspondent from Chester-le-Street described a chaldron waggon like the ones used on the Throckley Way. He records how these steep descents were tackled using the convoy.

> 'A waggon has four wheels. The two fore wheels are cast of metal for that purpose, and weigh several hundredweight as represented at C.D; the two hind wheels are of wood; the extremities of the axles are fixed in the wheels and turn round; the axles are iron; at F is a convoy, so called by the waggon-men but it may more properly be called a lever, for it is by this the waggon is guided down runs, or what may be called a precipice, or bank... was it not for this convoy, or lever, it would be impossible to guide the motion of the waggon on runs... This convoy or lever is taken out of what the waggon-men call a convoy band, as at G; and then the convoy or lever presses upon the hind wheel as at H; at I is a loyter pin, so called by the waggon-men, which pin is put into the end of the

---

[137] NEIMME: WBLB I p. 258 William Brown to Carlyle Spedding, 30th July 1754.

convoy, or lever, to hold it so as it may not jump out of the iron ring at K; sometimes they have pieces of small wood, or what may be called wedges, they put into the ring, to keep the convoy tighter, which the waggon-men call scotches, and they lie in what they call a scotch-box, as represented at L; the driver or waggon-man always has the convoy in readiness against he comes to the top of those runs, or banks, and then instantly jumps upon the convoy, or lever, and so by his weight and strength pressing upon the convoy to stop the waggon as he thinks proper, so as to let the waggon go fast or slow, till he gets down such places; they commonly unloose the horse when they come to the runs, and then put him too again when down; the reason of that taking him off at such places is, because, were the convoy to break, it would be impossible to save the horse from being killed, or if the waggon-way-rails be wet sometimes a man cannot stop the waggon with the convoy; and where the convoy presses upon the wheel it will fire and flame surprisingly; many are the accidents that have happened as aforesaid; many hundreds of poor people and horses have lost their lives; for was there ever so many waggons before the waggon that breaks its convoy and has not got quite clear of the run, they are in a great danger, both men and horses, of being killed.'

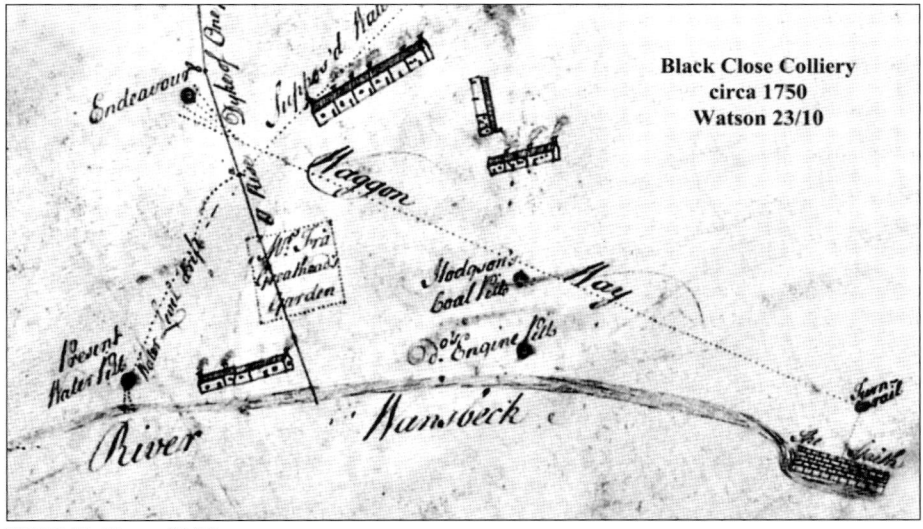

NEIMME: Watson 23/10

Fig. 33: Black Close Colliery circa 1750

Another safety device was the turnrail, such as the one shown on Isaac Thompson's plan of Newburn serving Humble's colliery (fig.7), which controlled access to the staith.[138] Waggons would need to be under control to be turned in the direction of the staith, a large, highly inflammable wooden structure. If they were not under control, or if they were on fire, they would simply run ahead away from the staith. The plan of the Duke of Portland's Black Close Colliery near Ashington (fig.33), with which William Brown was associated, illustrates the same point. In a letter to William Brown in 1755, Carlisle Spedding described an alternative method using a man to divert wayward waggons into a branch line by means of a switch rail.[139]

---

[138] William Hutchinson, a contemporary observer, records in his View of Northumberland (1776), Volume 2, page 417 that 'the carriages on an easy descent run without horses and sometimes with that rapidity, that a piece of wood, called a tiller, is obliged to be applied to one wheel and pressed thereon by the weight of the attendant who sits on it, to retard the motion: by the friction of which frequently the tiller and sometimes the carriage is set on fire'.

[139] NEIMME: WBLB William Brown to Carlisle Spedding, 15th June 1755, p.264.

From the wash hole the waggonway ran down to Newburn village. The map of Walbottle Moor Colliery (fig.34) shows the fifty five properties which were occupied principally by miners and keelmen. The Wylam Way ran between Water Row and the Manor House parallel to the main street and the stone bridge which carried it across the burn survives. The Throckley Way is off the map beyond the east side of the burn.

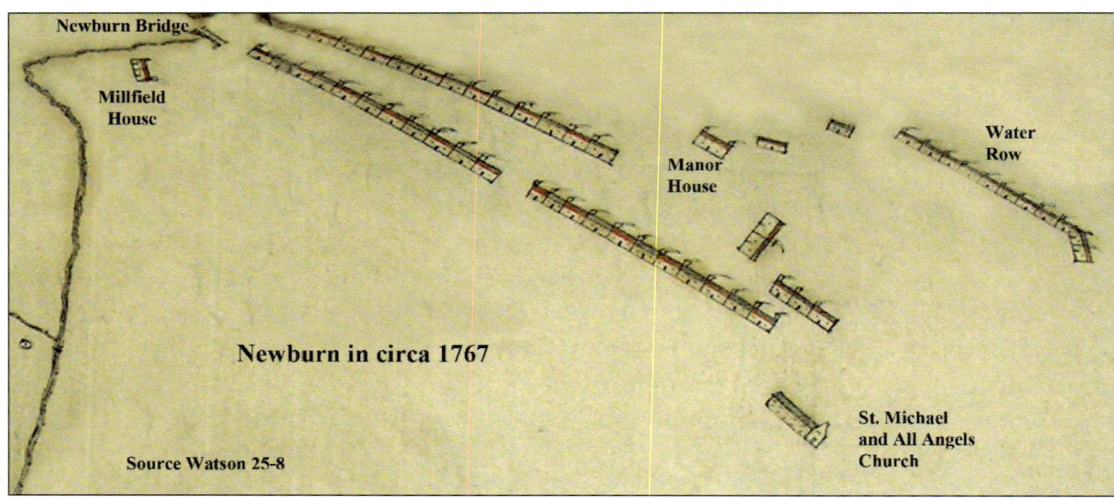

NEIMME: Watson 25/8

Fig. 34: Newburn Village circa 1767

At Falcus's garden, to the east of Newburn Bridge, a major change took place as a result of Brown's redevelopment. Originally, both the main and bye ways of the Throckley Way ran along the upper route (cccc). As part of the redevelopment, a new main way was built lower down for the partnership's line (nnn - lll) and the old Throckley Way became the partnership's bye way (fig.25). The Wylam Way, which ran between partnership's main way and bye way, is not marked. This could be because the colliery was flooded when the survey was made following the great storm of November 1771; or simply because the line belonged to John Blackett who was not part of the partnership. The Wylam Way crossed Brown's main way and bye way diagonally on what is later railway parlance would be known as diamond crossings.

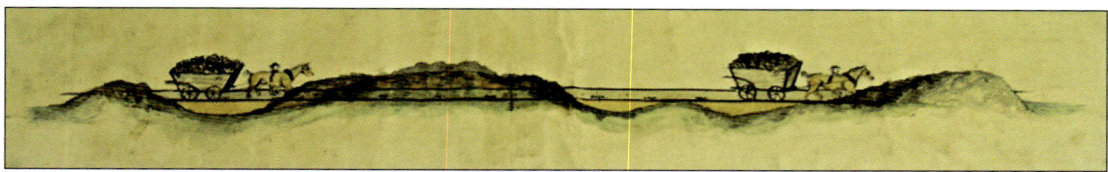

Courtesy of Tyne and Wear Archives and Museums

Fig. 35: The Waggonmen

'We have been laying down our New Waggon Way from Whingill this 5 Weeks and manage the Waggons pretty well in that Run with our two convoys on the fore Wheels. When the Rails are wett we lay small freestones about the bigness of a hen's egg which crush to bite upon the Railes and makes the Convoy take more hold of the Metal Wheels so that we can do pritty well in wet weather. But to prevent great mischief by the Running of Waggons, I intend at the bottom of the Run to have a Branch Laid and a Switch Rail to be shut by an Old Man to sit in a cabin. When he sees a Waggon got loose, he may shut the Switch Rails and turn the Waggon into the Branch which, running till it runs against the Battery of Earth and Turf, will take no harm and will fall back again gently. I'll have another Switch Rail which shall Shutt of itself and turn the Waggon (in its coming back) into another branch where it will rest until brot into the Main Way'.

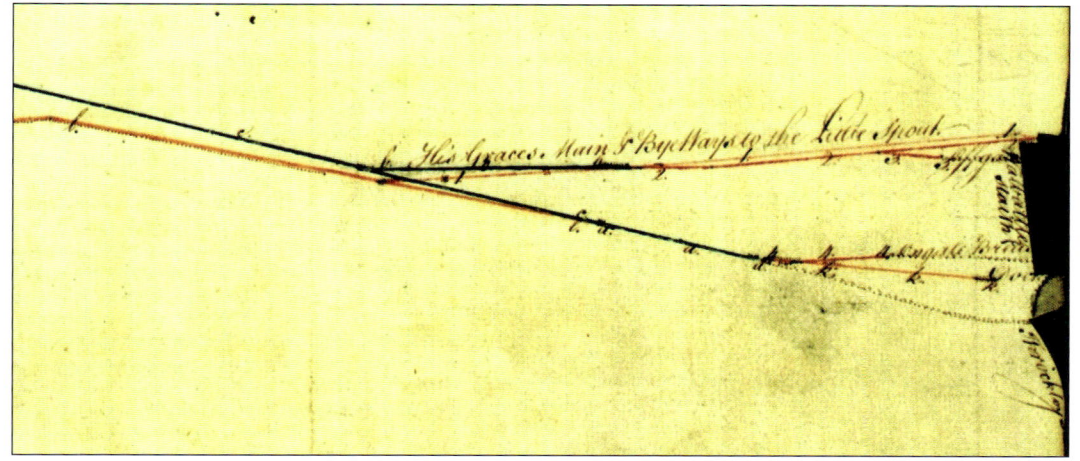

Courtesy of the Society of Antiquaries of Newcastle upon Tyne: NRO: SANT-BEQ-09-01-01-31
Fig. 36: The Throckley and Walbottle Staithes at Lemington

At the terminus of the line was the little spout, Walbottle staith, a dock and Throckley staith. A section of double main way (aaa) led to Throckley staith and it was also used by the partnership to access the dock (kkk) and by Walbottle Colliery to access the ongate to the staith (444). A main way (111) ran from the partnership main way ran to the little spout; and a bye way (222) ran from the spout to the partnership bye way. The short branch (333) connected the offgate from Walbottle staith to the bye way.

The community at Lemington in 1771 consisted of Walbottle staith house which had eight rooms plus a drinking room and cellar; and next door was what is described as a 'necessary house'. Humble's old engine house had been converted into a tenement of ten rooms for the keelmen. Adjoining was a house of four rooms rented by Richardson 'the staithman for the owners of Halliwell Rheins'. The 'little house on the bank going up to Wylam staith', which was formerly a smiths shop built by Humble, was in 1771 used by Cramlington. Wylam staith house and Throckley staith house, together with two cottages made up the other buildings.[140]

The original waggoners bond for Walbottle Colliery, dated March 17th 1770, survives.[141] It records the names of the persons supplying the waggons and the conditions under which they were employed. There were ten waggons engaged in total: three belonged to Henry Forster, two to Ralph Davison and one each to William Bell, William Longridge, Mary Hudson, George Dobson and William Westmorland. They were 'to be paid … the sum of seven and a half pence for each waggon Load of coals they shall respectively fill and lead from any of the pitts…to the Staith at Lemington such sum of Money as shall become due to be paid once in each half Year'. The waggonway provided these tenants of the Duke with an important additional source of income.[142] Given that the round trip was less than four miles, the drivers could have easily completed five trips in a day. The production of the colliery was in the region of 13,000 chaldrons each year which provided them with potential earnings of about £32 annually, at least half of which would be needed for the upkeep of the horse. The waggoners were required to 'obey the commands of the Staithman…as to the Mode of Filling the waggons and everything incumbent upon them as Waggonmen'. The

---

[140] The archives of the Duke of Northumberland at Alnwick Castle K/I/17/i/9/8
[141] NEIMME: Watson 2/13 – loose document at end.
[142] See Les Turnbull, 'Railways before George Stephenson' (Oxford 2012), Chapter Five which discusses the role of the tenants on Ridley's estates as waggonmen on the Plessey Way.

staithman was the person in charge of the operation of the waggonway: he had the delicate job of balancing the delivery from the colliery with the demands of the captains of the collier ships. The waggonmen were using a 24 Bolls waggon which was slightly more than a chaldron; and it is interesting to note that 'the measure of the waggons' was 'to be a Newcastle Chalder at least when disloaded at the Staith'. As with the miners, infringement of the bond incurred a penalty of £50.

Some details of the operation of the waggonway are revealed in a letter from the Walbottle staithman, Thomas Taylor, to the Duke's agent in November 1775. It shows that the waggonmen only worked during daylight hours and consequently more waggonmen were needed during the short days of winter. At this time there were thirteen waggons in use and it was proposed to increase that number to fifteen.

> 'I have the pleasure to inform you that we get very well forward at present in working coals, and likewise that we vend them as fast as they are had, not having one Calder of Round on the Staith – But as the days are now very short and we must expect that the weather will be bad. Consequently the Waggonmen Cannot go so many gates as they had in Summer, therefore it will Require two more waggons than we now have going to perform the same work the others did. Should be glad of your Directions – whom of his Graces Tenants I must Call upon to Set them on, as I plainly see we cannot Lead the Coals this Winter without two more and you know what a great loss His Grace Sustains if the Coals are not taken away as wrought…..It would be necessary to have fifteen waggons on till March and then we could reduce them again. I shall expect your directions as soon as possible'.[143]

Whilst the overall management of the railway was under the control of the viewer, the day to day maintenance was in the hands of a contractor and the day to day control of traffic was the responsibility of the staithman. The traffic down to the staith was predominantly coal for the seasale trade but certain railways did transport other traffic when they were able to take advantage of special opportunities: the Chopwell Way carried argentiferous lead pigs for the Blackett's refinery at Blaydon from Leadgate to Stella; the Whitley Way carried limestone from Marden Quarry and ironstone from Whitley Links to North Shields; grindstones were carried from Windy Nook to Felling staith on the North Birtley Way. Return traffic from the staith, such as lime and manure for the farms and bricks and wood for the colliery, was small in comparison but not insignificant: between 1824 and 1826 an average of over 1,000 tons of manure, which had arrived as ballast in the collier ships returning to the Tyne, was shipped to the farms along the Willington Way. Swedish bar iron was carried up the Derwent Way to Crowley's iron works at Winlaton Mill. As has been shown in chapter three, the vend and the wayleave returns provide a reasonably accurate figure for the traffic in seasale coal down to the staiths but there are few statistics available for the return traffic. In 1778, the Throckley and Walbottle Way transported 27,832 chaldrons of seasale coal made up of 14,686 journeys from Throckley Colliery and 13,186 from Walbottle Colliery. Given that there was a maximum of 275 working days, the average volume of traffic on the line was 101 waggons a day in each direction. The waggonmen worked a maximum of twelve hours each day and therefore there was a waggon passing each way every seven minutes. However, there were much busier lines: the Bigges Main Way

---

[143] The archives of the Duke of Northumberland at Alnwick Castle K17/i/9/7

carried twice this volume of traffic from Bigges Main and Willington collieries, amounting to 60,000 chaldrons per annum or on average 218 waggon in each direction per day or at least one each way every three minutes. A railway with a train every four minutes is a busy line and the maintenance crew would have a difficult task accessing the line for repairs; their problems would increase if there was a waggon every minute and a half.

For two centuries, before the advent of steam engines and inclined planes, horses were the motive power on the waggonways. Indeed, horses were used in large numbers after the development of steam power: in 1914 the eleven largest English railway companies were still using nearly 26,000 horses and when these railways were nationalised in 1948, British Rail acquired 9,193 horses which was a little less than half the number of steam engines.[144] The topography of the coalfield enabled most waggonways to be designed with a gentle falling gradient from the pits to the riverside which provided considerable assistance to a horse hauling a load of about four tons. There were some notable exceptions to this general rule, such as the steep climb to Blakelaw from Kenton Bank Foot on the North Brunton Way, and it maybe that banking horses were employed as they were on the roads. Horses were not only the principal source of power on the waggonway: they were deployed in many capacities throughout the colliery. However, like the steam locomotives which replaced them, horses needed fuel and maintenance; and these costs, especially of the cost of their feed, were major items in the budget of the colliery. At Bigges Main Colliery, for example, the cost of hay alone was generally three times the maintenance bill for the whole waggonway. Indeed, it was the rising cost of this feed in the early nineteenth century together with the increased volume of traffic which were the key factors stimulating engineers to experiment with iron rails, steam engines and inclined planes.

It is particularly interesting to compare the cost of using horses against using steam locomotives. In November 1812, John Blenkinsop provided John Watson with an estimate of the difference for the Kenton and Coxlodge Waggonway.[145] There were four elements to the cost of using horses on the waggonway. Firstly, the cost of feed, which was about 75% of the overall costs. It was estimated that four tons of hay, 104 bushels of oats, 26 bushels of beans and 15 weeks of summer grazing was needed by

---

[144] Biddle and Simons, 'Oxford Companion to British Railway History' (Oxford, 2000) p.212.

[145] NEIMME: Buddle 3/28, 24/46, 55/26; Watson 3/112/6a and especially Watson 3/13/108

Food for 81 Horses at £50 p. annum each	£4,050 – 00 – 0	
Trapping for 81 Horses at 3 Gns each	£255 – 03 – 0	
Farriers Wages, Drugs and Shoeing for 81 Horses at 1s each Horse p week	£210 – 12 – 0	
Decay of Horses, say 8 at £45 each	£360 – 00 – 0	
Creasing and Ballasting the Road 5½ miles at £150 p mile	£825 – 00 – 0	
Waggonmens wages for driving on 60,000 Cha. at 1s	£3,000 – 00 – 0	
8 Horsekeepers Wages at 18/- p week each	£374 – 08 – 0	
House Rent & Fire Coal for 81 Waggoners & 8 Horsekeepers at 2/6 p week each	£578 – 10 – 0	
		£9,653 – 13 – 0
Deduct for manure	£200	
Total expense		**£9,453 – 13 – 0**
Expense of conveying coals by Patent Steam Engine	£1,458 – 04 – 0	
Savings to the Owners		**£7,995 – 09 – 0**

each horse. Secondly, since a horse was only expected to last about five years, the deterioration in the value of the animal as a result of its workload was another factor which amounted to about 10% of the overall costs. Shoeing, the wear and tear on the trappings and the cost of veterinary services amounted to about 5% of the costs. The remaining 10% was the expense of a horse keeper including his rent and free coals. There are many other points of interest in this document but two deserve special attention. Firstly, the cost of creasing and ballasting the railway for horse haulage which was considerable: this cost was in effect a statement of the damage which the horses did to the track and therefore it does not appear on the estimate for using locomotives. Secondly, Blenkinsop's plan for using locomotives, which envisaged using five locomotives along the entire route from Coxlodge to Carville, did not dispense with horses completely since four horses were still needed for shunting.

Such was the nature of the horse drawn, wooden waggonways of the eighteenth century, those busy and important links between the colliery and the staith, which were built and operated by a specialist workforce, Britain's first railwaymen. Like the Throckley and Walbottle waggonway, the network of lines to the principal rivers in the region shown on John Gibson's map, were complex railways which provided the essential transport facility for the collieries they served. Four different types of rail were used to meet the different circumstances along the line. The heavy double main way for loaded waggons and the lighter bye way for returning traffic were the principal elements but there was also an assortment of sidings and branches as on a modern railway. The staiths and the large bridges, such as the one across Walbottle Dene, were major engineering achievements as were the great cuttings and embankments like the mighty earthworks on the Tanfield line which attracted the attention of the great antiquarian Dr. William Stukeley. The wash hole, which provided an essential watering facility so necessary to protect and rejuvenate the wooden wheels and brakes of the waggons, was a special feature which has a particular interest because of the recent discoveries at Carville on William Brown's Willington Way. Certainly, these waggonways cannot be dismissed as a simple tramroads for they had many of the complexities of the iron railway of the nineteenth century; and, like the iron railways of Victoria's reign, they made possible a massive expansion of the coal industry which not only brought prosperity to the mining communities of Northumberland and Durham but changed the economic history of the country at large.

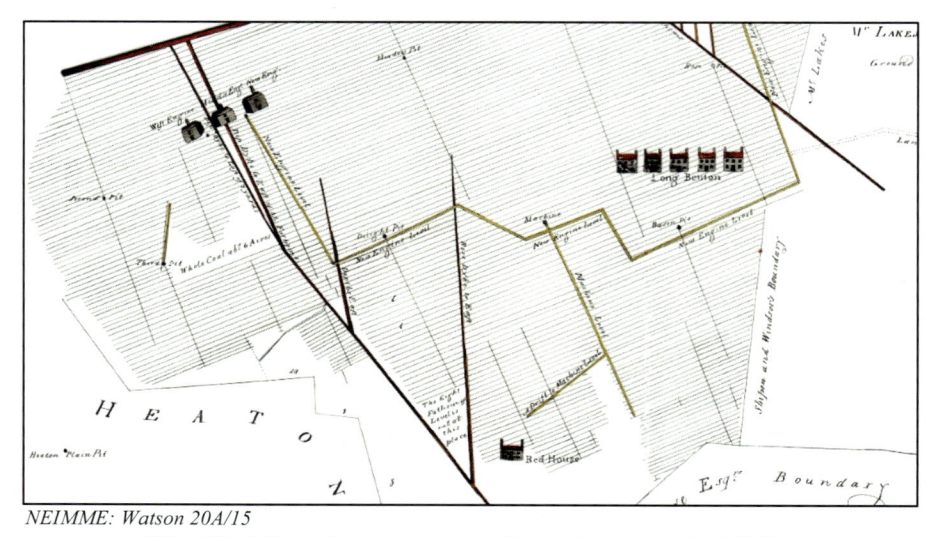

NEIMME: Watson 20A/15

Fig. 37: West Longbenton Fire Engines in 1762

Chapter Seven: Steam Engines – William Brown's Forte

Since steam engines were first invented, their size and power have attracted interest from women as well as boys of all ages. In June 1766, Mrs. Montagu wrote excitedly to her friend, the poetess and fellow bluestocking Elizabeth Carter, about her visit to East Denton Colliery where William Brown had built a massive pumping engine with a 60 inch cylinder:

> 'I went last night for the first time down to the Colliery, it lies on the edge of the Tyne…from the top of the Engine arises a vast column of black smoak and a cloud of steam at ye bottom issues out. This engine pours into the Tyne about 1,400 hogsheads of ill sented water in an hour to the great disgust of the River God and water nymphs'.[146]

In this letter she expressed the same thrill at seeing a large steam engine at work that had excited Sir John Clerk, Dr. William Stuckley and the Earl of Oxford earlier in the century; and still excites enthusiasts today. In 1766, across the fields to the east, Brown's assistant, Christopher Bedlington, was supervising the building of a more powerful engine in Benwell with a 75" cylinder weighing over five tons.[147] It was supplied with steam from three boilers and disgorged water into the Tyne through 24 inch pumps. To the west, Brown's men were sinking bore holes in preparation for the development of the Duke of Northumberland's colliery at Walbottle, where another huge engine with a 70" cylinder would soon be at work. Indeed, the name of William Brown is associated with many of the steam engines built in the Great Northern Coalfield between 1755 and his death in 1782. During this time, probably his greatest achievement was to win the deep coal of the Tyne Basin and make the area to the east of the Ouse Burn the principal source of Tyneside coal in the later eighteenth century.

Removing water from the mine was, and still is, one of the major problems facing mining engineers. So great was this problem that the weight of water to be lifted often exceeded the weight of coal, in some mines by a factor of seven or eight.[148] Where the landscape was favourable, drainage levels were driven to allow the water to flow out of the workings naturally; but this was not possible everywhere and, as mines became deeper, other solutions had to be found. Then horses, water power and even windmills were used to raise water either to a drainage level or to the top of the shaft. There were some spectacular examples of the use of water power for drainage in the region, notably Sir Henry Liddell's Ravensworth engine of 1670, which by means of three waterwheels drained of the western side of the Team Valley; and there were other major installations at Winlaton, Stella, Allerdene, Felling and Lumley.[149] Nonetheless, despite the development of the more efficient bob gin in about 1690, by the beginning of the

---

[146] Huntington Library MO 3174, Elizabeth Montagu to Elizabeth Carter 27th June 1766.

[147] Benwell royalty been leased to the Wortley branch of the Montagu family, founder members of the Grand Allies, in the early eighteenth century but was abandoned in the 1720's because of flooding. In all likelihood the Grand Allies kept the colliery as part of their portfolio and they were responsible for financing the large engine of 1765. The stages by which this engine was built are recorded in Bedlington's view book – NEIMME: East 1b; a summary appear in Raistrick's paper Trans. Newcomen Society Vol.17 p.149-52

[148] Taylor, J.T., 'Archeaology of the Coal Trade', 1852, p 191. He noted that 'in particular areas such as that of Percy Main Colliery and Wylam Colliery, and in other instances, it is not unusually seven or eight times' the amount of coal.

[149] Eric Clavering, 'Coalmills in Tyne and Wear Collieries', Bulletin of the Peak District Mines Historical Society, Vol.12, No.3, Summer 1994, provides a detailed analysis of the use of the waterwheel for mine drainage between 1600 and 1750.

eighteenth century, the coal industry in Northumberland and Durham was facing a crisis because much of the coal was inaccessible since it lay at a depth which was beyond the capacity of the pumping machinery of the day to drain.

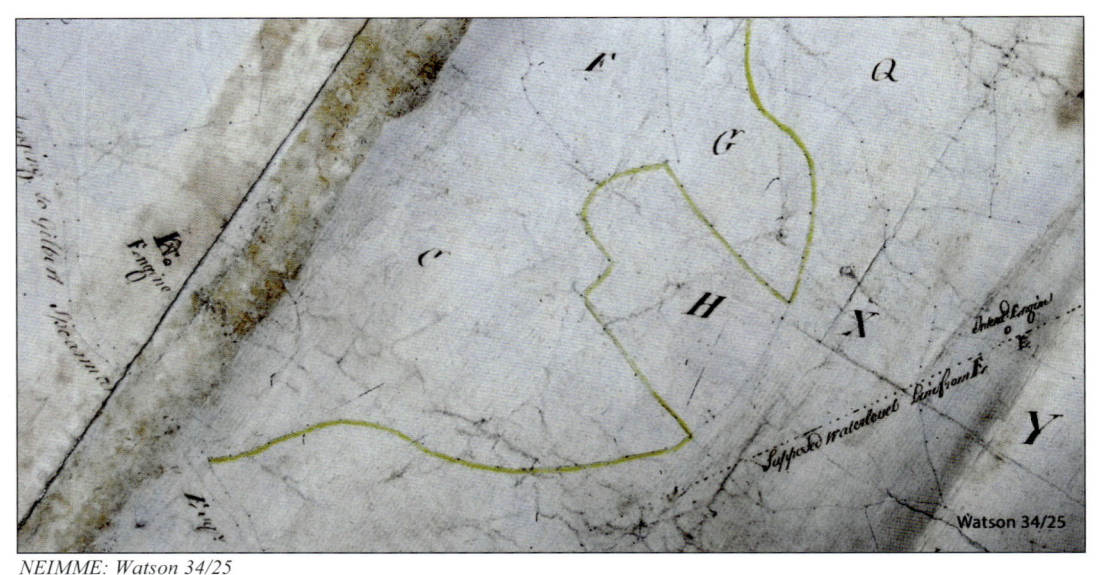

NEIMME: Watson 34/25

Fig. 38: A Newcomen Engine at Tanfield Lea in 1715

Throughout the seventeenth century men of science, such as Robert Boyle and his friend Denis Papin, were intrigued by the power of air and steam. Their investigations culminated in Thomas Savery's demonstration of a pumping engine to The Royal Society in 1699, which he later described in detail in his book 'The Miner's Friend; or An Engine to Raise Water by Fire'. Unfortunately, his engine could only raise water thirty feet and consequently it was no solution to the problems of the deep mines of Tyneside. However, during a decade of experiment, Thomas Newcomen improved upon Savery's invention and by 1712 he had developed a successful engine which was released under Savery's patent. The engine was immediately adopted in the Great Northern Coalfield. The map of Gilbert Spearman's estate at Tanfield Lea in 1715 (fig.38) shows one of the first Newcomen fire engines in the northern coalfield with its large chimney, coffee-pot boiler and regulator beam above the pit shaft. The map also records that a second engine was being planned.[150] By 1720, similar engines had been built for collieries at Oxclose on Washington Fell, at Norwood at the head of the Team Valley, at Byker to the east and Elswick to the west of Newcastle. Such was the importance of this new pumping technology that Robert Galloway, an early authority on the history of the mining industry, formed the judgement that 'the invention of the steam engine may safely be said to have been the most important event that has ever happened in the annals of mining'.[151]

When Thomas Savery died in 1715, a syndicate of London speculators was appointed to promote the Newcomen engine. In 1718, an agreement was made with Richard Ridley and by the time of Sir John Clerk's visit in 1724 three engines were working at

---

[150] There is no evidence that the second engine was built but a map dated 1750 (Watson 31/14) of Tanfield Moor Edge and Bushblades collieries shows a replacement engine for the first. In 1729, Richard Ridley acquired the estate and the engine on Watson's plan is probably the 1730 replacement (NEIMME: Forster 1/5/39). It is tempting to suggest, because of his enthusiasm for engines at Byker, that Ridley was the man responsible for the first engine at Tanfield Lea. See appendix.

[151] R.L. Gallaway, Annals of Coalmining and the Coal Trade, London, 1898, p. 236.

Byker Colliery. Probably similar arrangements were made with the other major coalowners. Certainly, William Coatsworth and Henry Liddell were in correspondence with the syndicate for engines at Gateshead Park, Farnacres and Heaton Banks collieries. However, their relationship was a fractious one and although engine houses were built no engines were delivered immediately to Park and Farnaces; and an engine was not at work at Heaton Banks Colliery until April 1729. No doubt the shortage of engine parts and skilled labour to build and operate this new technology could explain in part the delay; but there were more potent reasons. Cotesworth correspondence indicates that the syndicate were quite ruthless in their business dealings and that they were open to bribes from Ridley.[152] The contrast between Ridley's success and Liddell's initial failure was doubtless another dimension to the rivalry between these two factions in the northern coalfield which culminated in 1725 in Ridley's attempt to have Liddell's partner, Cotesworth, poisoned. Despite this friction, by the mid 1730's the greatest concentration of steam engines in the world was to be found in the lower reaches of the Ouse Burn Valley on the edge of the Tyne Basin. Besides the three engines at Byker, there was at least another three at Jesmond; and at Heaton Banks Colliery, belonging to the Grand Allies, four engines were constructed between 1729 and 1732. A fifth engine was added in 1733 but this was a water powered bob gin not a fire engine. The location of the fire engines in Heaton and Jesmond are shown in fig.39.[153]

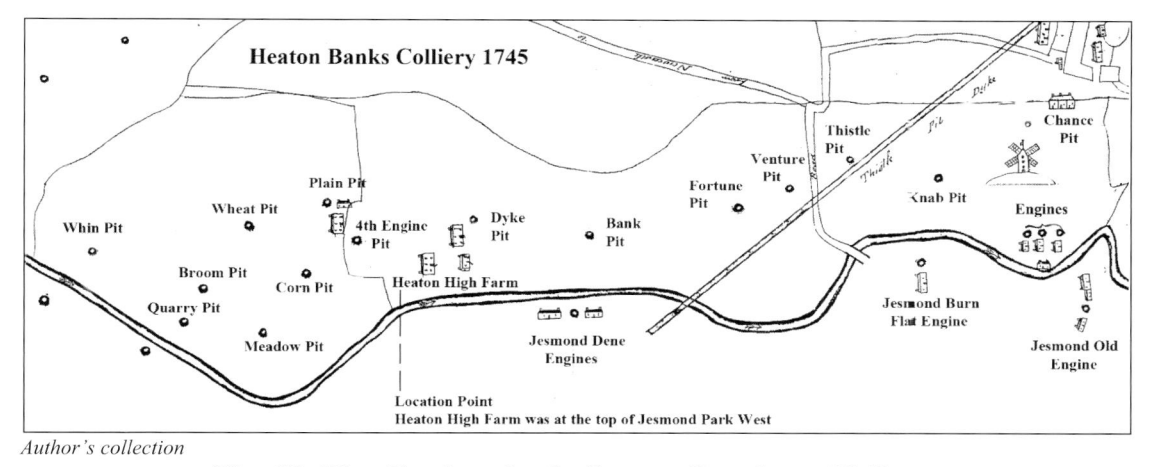

Author's collection

Fig. 39: Fire Engines in the Lower Ouseburn Valley.

Although John Potter of Chester-le-Street was appointed the syndicate's agent in the north by 1724, his principal role seems to have been to manage the engines at Flatt's and Newbottle collieries. The work of building the engines appears to have been carried out by local viewers such as Amos Barnes and Richard Peck. The steam engine was also quickly adopted by the collieries shipping from the River Wear. Henry Beighton had built an engine for Oxclose Colliery, Washington, by 1718; and the engine at Flatts Colliery was in operation by 1723. Henry Lambton had also experienced difficulties in securing an engine from the syndicate for Biddick. However, Burleigh and Thompson's plan of the River Wear in 1737 (fig.40) shows that engines had been installed at North Biddick (7), South Biddick (13), Chartershaugh (3), and on esquire Lambton's home estate (10). Lord Scarborough of Lumley Castle had an engine installed at Newbottle Colliery in 1733; and by the following year, the Newbottle engine had become a pawn

---

[152] Archaeologia Aeliana 4th series 27 (1949) Edward Hughes 'The First Steam Engines in the Durham Coalfield'.

[153] The map is based upon a plan of Heaton Estate in Northumberland Record Office ZRI 50/9.

in an industrial dispute between John Nesham, the new leasee, and the pitmen concerning the size of the corves in use at the pit. The conflict had resulted in the accidental shooting of one of the miners. In the court case that followed, the engineman at Newbottle, the syndicate's man John Potter, testified that 100 pitmen 'threatened to murder him and pull down the engine if he set her to work again'.[154] By the early 1740's another engine had been built for Lord Scarborough at Lumley Colliery.

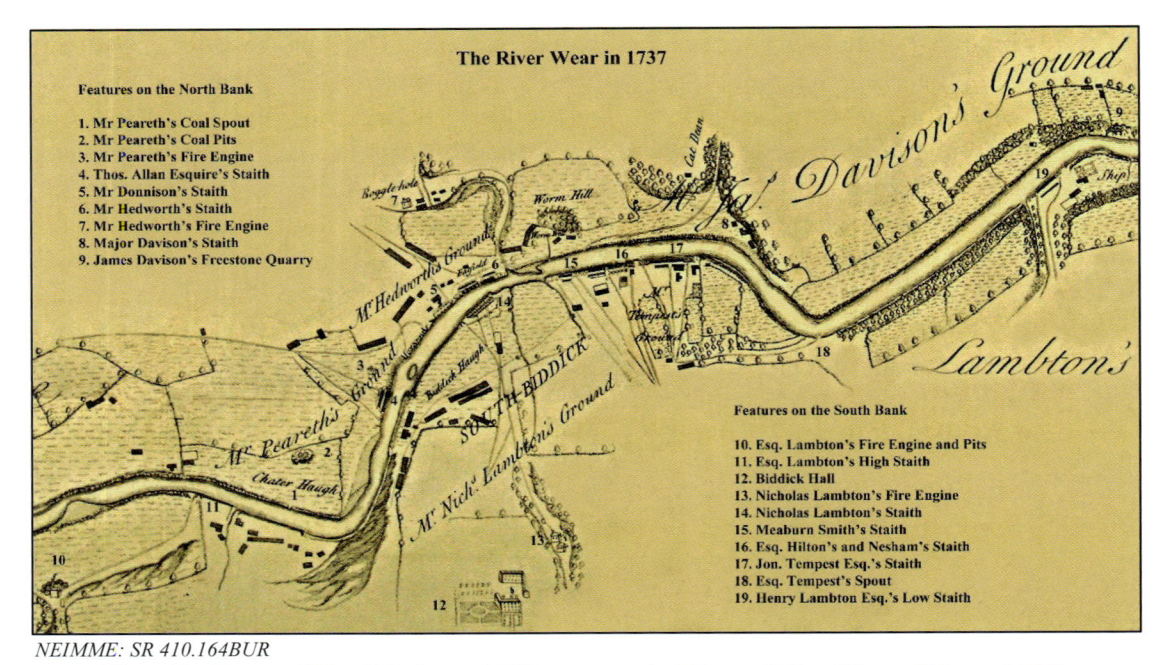

NEIMME: SR 410.164BUR

Fig. 40: Part of Burleigh and Thompson's Plan of the River Wear 1737

The Newcomen engine worked by condensing steam in a cylinder by means of an injection of water and the resulting vacuum enabled the air pressure to move the piston. The principal parts of the engine were the boiler, the cylinder, the piston, the regulator beam, the water injection apparatus and the pumps. A quotation given in 1733 by Richard Peck for work at both Jesmond and Byker collieries reveals that the total cost of an engine was £849: a cylinder cost £150, a boiler £112 and an engine house £120. There were other ancillary expenses such as driving levels, sinking shafts and providing water courses. In 1730, the engine to serve Bushblades Colliery on Tanfield Moor cost £800 but driving the drainage level 1,200 yards from Tanfield Lea cost a further £1,080; at Jarrow Colliery, in 1742, the engine and engine house at High Heworth cost £1,200 and sinking the engine pit cost at additional £522.[155] Like the waggonway, the pumping engine was a substantial part of the capital costs of winning a colliery and only the larger enterprises could afford this new technology.

The view book of Amos Barnes, the colliery engineer at Heaton, contains a diagram of an engine house and one of the three engines in Low Heaton which raised water into the Tyne Level Drift (fig.41). It is interesting that after the four Newcomen engines finally arrived at Heaton a bob gin was installed underground in 1733 to tackle a particular drainage problem, a reminder that the old and new technologies existed side by side. The Newcomen engine is shown operating two sets of pumps: one set 'A', lifted water to the Tyne Level Drift and the other 'B' raised water to drive the wheel of

---

[154] Archaeologia Aelians 2nd series 2 (1858) 'Report on the Pitman's Strike at Newbottle in 1734'.
[155] NEIMME: Peck I p.73; Forster 1/5/39, 1/5/140; 1/4/180

the bob gin. The fire engine had a 42'' cylinder and a piston with a seven foot stroke: it was capable of ten strokes a minute and drew 220 hogsheads (11,550 galls) per hour. This was impressive for the day but later Brown would build engines with five times that power. The water engine had a four foot stroke: it was capable of six strokes a minute and pumped 62 hogheads (3,255 galls) each hour.[156] However, difficulties with water continued to trouble the management and the dip of the colliery (the southern part) was abandoned. In an effort to keep the pits in the northern part working, the lift of the pumps was changed from 44 to 22 fathoms which increased the amount of water being discharged. Simultaneously, the Grand Allies began to develop West Longbenton Colliery, lying between the north of Heaton estate and the Ninety Fathom Fault, as a replacement. Heaton Banks Colliery lost its fight against water in 1745, the year West Longbenton opened. Peck's Colliery in Jesmond also closed in that year.

NEIMME: Forster 1/4/11

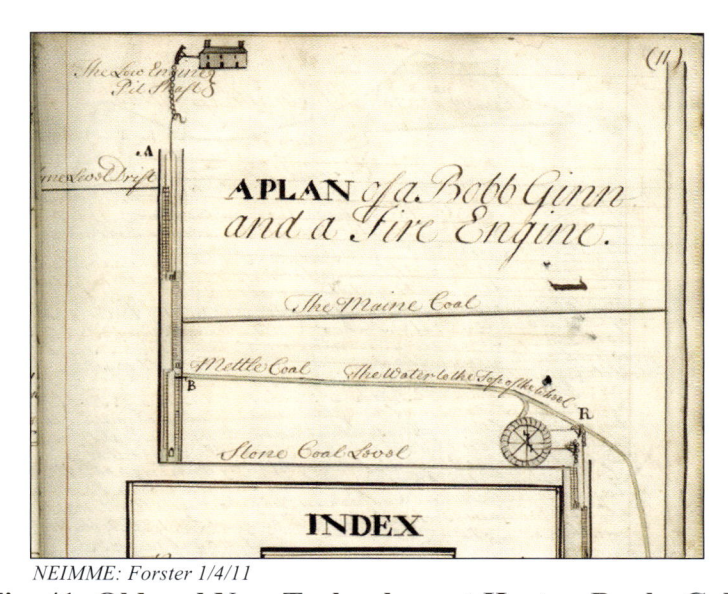

Fig. 41: Old and New Technology at Heaton Banks Colliery

As has been shown in Chapter Two, part of William Brown's talent was that he was able to recruit men with the necessary skills to complete these major engineering projects. Also, he had the connections to commission the various materials needed. The smaller brass cylinders of the early engines were probably produced locally; the larger iron cylinders used later by William Brown were mainly built by Abraham Darby's foundry at Coalbrookdale. Brown was closely associated with the Midlothian collieries and he would undoubtedly have had connections with the Carron Iron Works. Through his work at Hartley and Longbenton collieries, he knew John Smeaton, the engineer credited with establishing the boring mill at Carron. The wrought iron plates for the boilers were probably produced locally by Hawks of Gateshead or Crowley's crew at Swalwell, then the largest iron manufactory in Europe, which was within sight of Brown's house. There was a long tradition of supplying such plates to build large iron pans for evaporating salt from sea water which was a major industry in the region. Brown's correspondence with Leonard Hartley reveals that Yorkshire was one of his sources for the massive timbers needed for his projects.[157]

---

[156] NEIMME: Forster 1/4/11-13.

[157] The most comprehensive study of the tasks involved in building a Newcomen engine is to be found in Steve Grudgings paper in the Transactions of the 2014 NAHMO conference.

A plan of the Earl of Carlisle's estate dated 1749 drawn by John Watson shows the pumping engines at the First Pit (1) and Second Pit (2) supported by an engine further west just across the estate boundary in Gosforth royalty. A fourth engine at the Lane Pit (4) appears to be a winding engine. This is very significant for steam power is generally not thought to have been applied to winding until two decades later. As pits became deeper, the problem of winding coal to the surface increased and several engineers turned their attention to this issue. Both Jonathan Hull and John Wise took out patents for devices to transform the reciprocating motion of the Newcomen engine into rotary motion but nothing appears to have come of these inventions.[158] Joseph Oxley modified a Newcomen engine at Hartley which enabled the colliery to dispense with horses for winding; and this engine was seen by James Watt in 1768. Both the Longbenton and Hartley engines would have been known to William Brown. The winding engine at the Lane Pit is not marked on a map of Longbenton dated 1762 (fig.37) which suggests that the experiment was unsuccessful. However, a third pumping engine, described as the 'new engine', had been built to the east of the Second Pit.[159]

NEIMME: Watson 20A/12

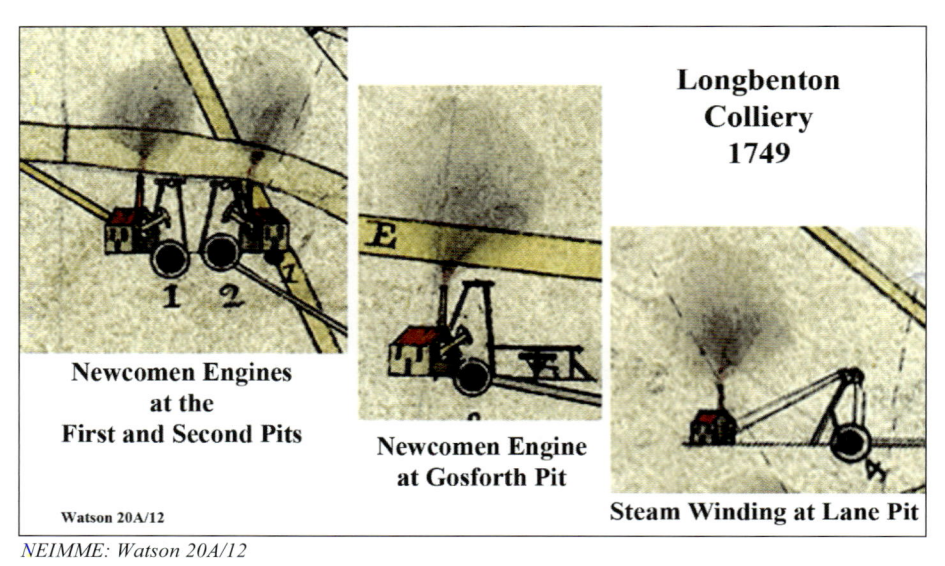

Fig. 42: Engines at West Longbenton Colliery in 1749

By 1762, West Longbenton Colliery was nearing exhaustion and plans were already well advanced for a move eastwards towards what would become known as Benton Square. In November 1760, the Grand Allies received an estimate for a 'Double Engine House', a 42 inch cylinder, 63 fathoms of 14 inch iron pumps, two engine pits, four coal pits, a smith's shop and a house together with 1,100 yards of ash and oak rail for the waggonway. The cost was £2,766 but not everything required is in the quotation: there is no mention of the other cylinder or of the boilers, which may well have been recycled from Longbenton West, and more timber was needed for the waggonway. William Brown's name has been linked with this job on the northern edge of the Tyne Basin: he appears to have negotiated the cylinders from Coalbrookdale. The name of John Smeaton is also associated with this colliery for he built a pumping engine to raise water to drive winding machinery at the Prosperous Pit in 1777.[160]

---

[158] NEIMME: Trans. M.I. Vol.82 p.529; International Journal for the History of Engineering and Technology Vol. 82 No. 2 (2012) p. 176-186 - Steve Gudgings 'John Wise – unrecognised engine builder and contemporary of Newcomen and Watt'.
[159] NEIMME: Watson 20A/9 and 20A/13.
[160] NEIMME: Forster 1/4/199; Transactions 113, Raistrick p.23; Transactions 82, Louis p.527.
International Journal for the History of Engineering and Technology Vol.82 No.2 July 2012 p.176-86, Steve Grudgings 'John Wise – Unrecognized Engine Builder and contemporary of Newcomen and Watt'.

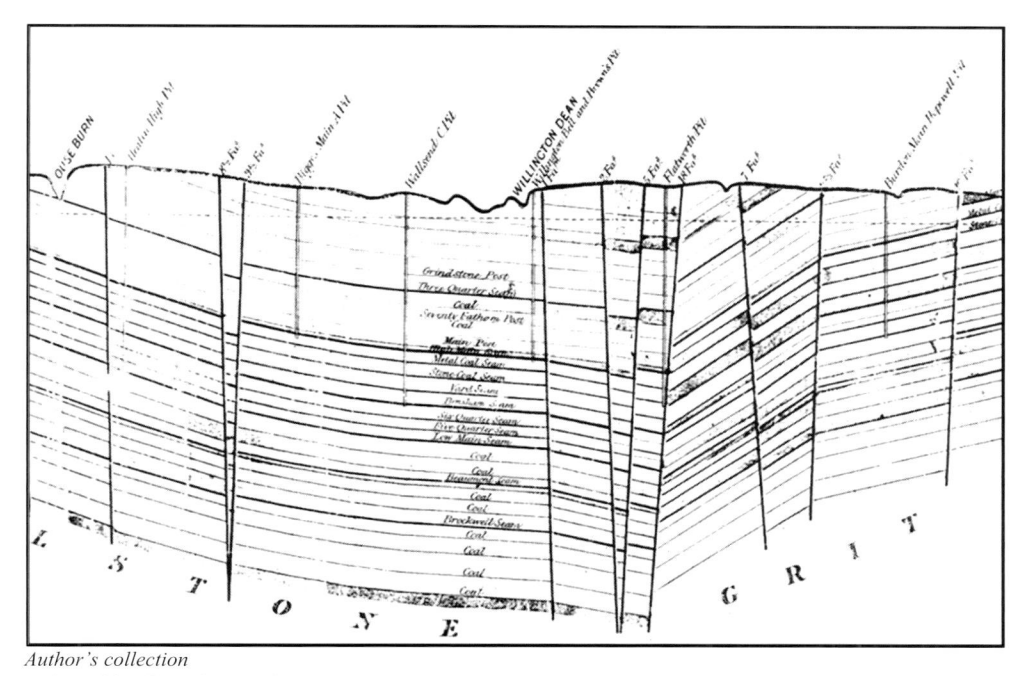

Author's collection
Fig. 43: Section of the Strata in the Tyne Basin by William Oliver 1861

William Oliver's map of the strata between the Ouseburn and Tynemouth (fig.43) shows the coal seams dipping from the west and the east to form the trough of the Tyne Basin. One of the major challenges to mining engineers in the northern coalfield during the second half of the eighteenth century was to win the precious High Main coal, a six foot seam of top quality household coal, from this extensive area. From the fragmentary evidence that survives, it is not possible to write a detailed account of the exploitation of the Tyne Basin; but it is clear that William Brown and his assistants, Christopher Bedlington, William Gibson, John Allen and George Johnson, were the men principally responsible. They undoubtedly had the support of John Watson who was living at Willington House. In 1750, apart from the collieries on the edge of the basin at Jesmond, Byker, Heaton Banks, Walker Hill and West Longbenton, all the deeper coal was inaccessible and many, like William Brown, thought it would remain so for the foreseeable future. In a letter to Leonard Hartley, concerning Humble's attempt to win Wallsend in 1752, he expressed these doubts.

> 'Twixt you and I Wallsend Colly should be the last in the Country I would give a farthing. For in the first place if there is a seam of coale at all that is regular she must be exceeding deep… she lies some Thousand yards to ye East of dip of Longbenton, Heaton and Byker again she lies a great distance to the west which is the dip of Billy Mill Colly so that it is hard to acc't in any tolerable way for her. Mr Humble has bored a very great hole in the Estate that joins her to the east and I know very well he has not yet got any workable seams of coale yet he is so odd a mortal that he will do things that is very uncommon. Therefor the Dean and Chapter would certainly do well to tye him fast to a high annual rent work or not work. That must be most to their advantage for believe me nobody now living will ever see her a working Colly'. [161]

---

[161] NEIMME: LBWB 1 p.101 William Brown to Leonard Hartley 31st October 1752.

Time would prove William Brown wrong and he would be personally instrumental in winning what was to become the greatest colliery in the coalfield, Wallsend, whose name would become a byword for top quality domestic fuel.[162] By the end of the century, the Tyne Basin would replace Tanfield Moor as the principal source of household coal as a consequence of William Brown's mastery of pumping technology. But in 1752, Brown's interest lay elsewhere – in Jarrow Colliery across the river. This was not the town of Jarrow but the large area of land lying to the east of the high road from Newcastle to Durham – what today we would refer to as Heworth, Hebburn and Jarrow. This colliery had been developed by Sir William Blackett and John Wilkinson first by using bob gins and then Newcomen engines. However, Brown's scheme came to nothing and his interest turned to introducing steam power to Throckley Colliery. On 11th July 1752, he wrote to his friend Carlisle Spedding who, helped by his elder brother John, had built the first Newcomen engine at Whitehaven between 1715 and 1717:

> 'thought of fixing a fire engine six hundred yards to the dip of our Tyne level…and resolv'd on to Take a Trip to your Place in order to begg a little of your advice as to the erecting of one of these machines.'[163]

However, it would be wrong to infer that William Brown was a novice at that time. He was the viewer at Humble's Newburn Colliery where a fire engine had been installed in 1749 by William Newton. The viewer and engine builder Richard Peck lived at Peck's Houses Newbiggin, about three miles to the north east of Brown's home at Throckley Pitt House and may well have been his mentor. In the close knit world of mining engineering, it is incredible that William Brown would not have been aware of the revolutionary work of his neighbour and the fourteen fire engines pumping at Byker, Jesmond, Heaton Banks and West Longbenton on the edge of the Tyne Basin. In January 1750, Brown was a junior member of a panel of seven viewers who reported upon West Longbenton Colliery which was threatened by flooding; and in 1752, he accompanied the distinguished viewers George Claughton, William Newton and Amos Barnes on an inspection of West Denton Colliery.[164] Both these collieries depended upon Newcomen pumping engines. Thus William Brown would be very familiar with pumping engines long before he travelled to Whitehaven to seek Spedding's advice.

In November 1752, William Newton, Amos Barnes and George Claughton made a view of Throckley and recommended that 'the Cylinder of the Fire Engine…be no less than Forty Two Inches Diameter' and expressed the opinion that 'when Completed will at Least win a Hundred Acres of Coal'. During the next thirty years, with the support of Isaac Thompson, the local representative for the Dale, William Brown was to build some of the largest pumping engines of the day. Transporting these cylinders from Shropshire to Tyneside was a major undertaking and it is likely that Thomas Goldney, the firm's agent in Bristol, played a major role. This heavy cargo would be taken down

---

[162] In 1816 Matthias Dunn recorded – 'It is not a little curious to note the number of Wallsends in the Coal List' – twelve collieries on the Tyne and two on the Wear used the prefix 'Wallsend' to market their coals. Newcastle Central Library: Viewbook of Matthias Dunn p.128.

[163] NEIMME: WBLB I p.68, William Brown to John Spedding, 11th July 1752.
[164] NEIMME: Forster 1/4/145, 222 and 231.

the River Severn on a barge and then transhipped to a collier at Bristol for the journey round the south coast to Tyneside.[165]

Scientific knowledge spread through personal contacts in the eighteenth century. In 1759, James Spedding, Carlisle's son, wrote to William Brown about the engine built for Thomas Broade's colliery by James Brindley at Fenton Vivian near Newcastle under Lyne. At the request of Lord Ravensworth and Sir John Hussey Delaval, their viewers Nicholas Walton and William Brown visited Brindley's engine later that year. However Brindley, who was heavily engaged on building the Duke of Bridgewater's canal, was unable to be present; and Mr Broade was reluctant to reveal all the secrets of the improvements, despite the letter of introduction which they carried. William wrote to Sir John that Mr Broade 'seem'd Willing we should see what may properly be called the External part of the Engine and pitts, but wd not let us see the inside of the Boiler'. However, the northerners had a strategy to overcome that problem: 'we soon learned how it was constructed by Dint of money we gave to the men to Drink and was master of the whole construction that Day'. The two viewers then travelled to Coalbrookdale where they saw another boiler of Brindley's design being built. Their status was such that they 'staid with Abraham Darby, he being an acquaintance of both Mr Walton and mine and master of the great iron Works there'. [166]

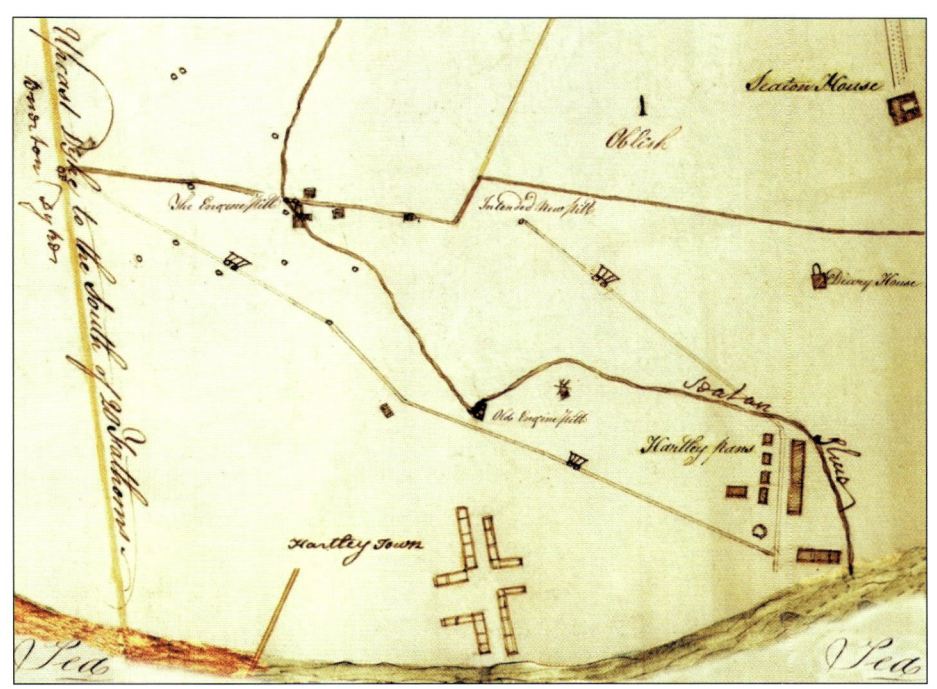

Courtesy of the Society of Antiquaries of Newcastle upon Tyne: NRO-SANT-BEQ-09-01-01-68
Fig. 44: Hartley and Seaton Delaval in the late 1750's

Sir Henry Liddell was anxious for the future of West Longbenton and his collieries at Ravensworth in the Team Valley; and Sir John Hussey Delaval was in the midst of

---

[165] Matthias Dunn (History of the Coal Trade, 1844, p41) is completely mistaken in writing that 'upon getting the management of Throckley colliery he built one there – then a great rarity'. See Ken Rogers, 'The Newcomen Engine in the West of England', which is based on the diary of Thomas Goldney the agent for Coalbrookdale at Bristol.

[166] NRO: 2DE 6/3/2 William Brown to Sir John Hussey Delaval, 3rd September 1759. The invitation to stay at Abraham Darby's house may be an outcome of their friendship with Isaac Thompson who was related by marriage to Abraham Darby. A full account of Brown's report including his drawing of the boiler, is available in Trans. Newcomen Society 76 (2006), Christine Richardson 'James Brindley (1716-72) – his Simultaneous Commercial Development of Mills, Steam Power and Canals'.

major developments at Hartley Colliery under the direction of William Brown, which included the erection of second pumping engine situated at the new winning further south. Improvements to the harbour at Seaton Sluice were also planned. John Smeaton's advice was sought and although he produced a plan his scheme was rejected. In December 1759, Brindley gave permission for one of his engines to be built at Seaton Delaval; and, on the 29th April 1761, Brown was able to inform Sir John that 'at last we got our great cylinder' from Coalbrookdale. However, the driving of the great cut through the sandstone headland to improve the harbour would take longer: the channel, which was 900 ft long, 30 ft broad and 52 feet deep and could accommodate twelve to fourteen vessels, was not opened until March 1764. Both the engine and the new cut contributed greatly to the expansion of trade from Seaton Sluice, which trebled in the next decade. This episode clearly indicates Brown's connections with the leading engineers of the day and it is a reminder that personal connections were a primary form of technology exchange in the eighteenth century.[167]

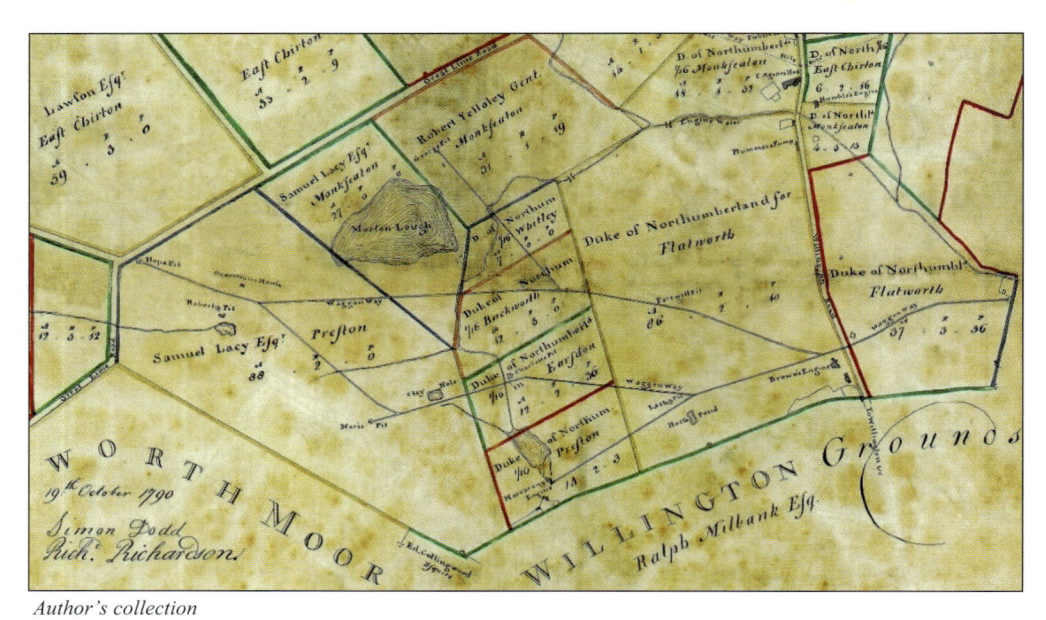

Author's collection

Fig. 45: Shiremoor Colliery in 1790

During the 1750's, William Brown was also involved in the development of Shiremoor Colliery on the eastern rim of the Tyne Basin. Since 1749, George Humble had been trying to win Shiremoor Colliery but had encountered major problems with water: all seven of the pits sunk in 1749 were lost to flooding. In 1755, Bell and Brown took shares in the lease for 21years and an engine was built. A decade later, in 1765, Brown and Bell bought Humble's share and they turned their attention to the deeper coal to the west of the moor. Mrs Montagu wrote to her husband Edward, in September 1767, that 'the D. of Northumberland is making Bell and Brown expend £10,000 upon a new winning at Shiremoor and if ye coal in ye new winning is no better than in ye old, as is supposed, the money will be entirely thrown away'. Much of this expense (£4,900) was on sinking the Engine Pit and the erection of a 'double engine house compleat with 72" and 60" cylinders and four boilers' similar to the one built at Walker Colliery in 1763. This became known as Brown's engine. In December 1767, Christopher Bedlington recorded the erection of 'one of the largest … Fire Engines now in use' to drain the lower part of the colliery; and The Newcastle Courant noted that 'this Engine will win

---

[167] Linsley, S., 'Ports and Harbours of Northumberland', Tempus 2005, p. 171-193.

or drain upward of a Million Chaldrons of Coal at a Depth of 70 Fathoms' and that it was 'capable to draw upwards of 1,000 Hogsheads (63,000 gallons) of Water per hour the above said depth'. The map of Shiremoor (fig.45) shows Humble's engines (top right) in the rise part of the mine and Brown's engine to the dip. The Marcarony engine further north was erected in the following year, 1768. Figure 45 also shows the branches of the waggonway which led south through Flatworth to Whitehill Point, only two miles from the mouth of the river, which was probably also Brown's handiwork.[168]

NEIMME: Watson 27/1

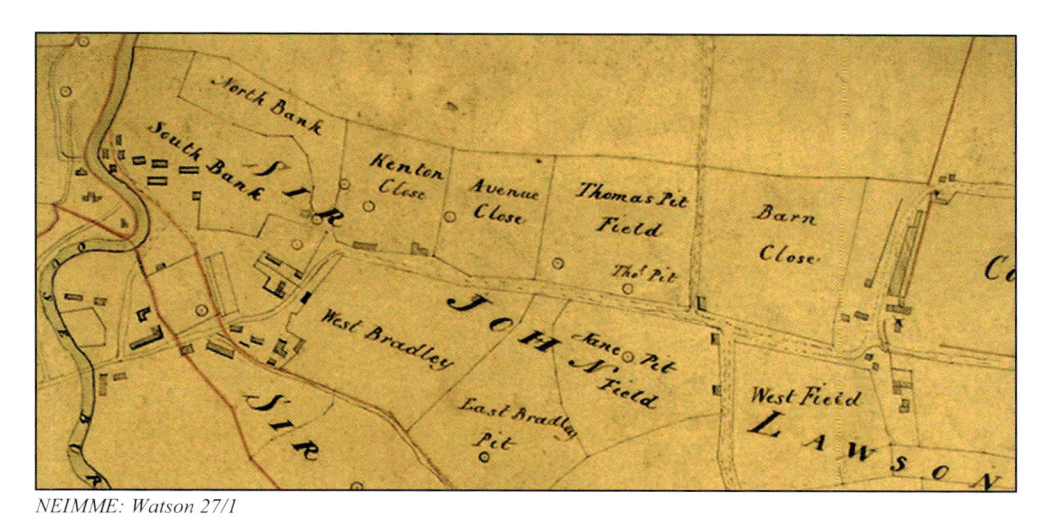

Fig. 46: Byker Old Colliery

NEIMME: Watson 25/19

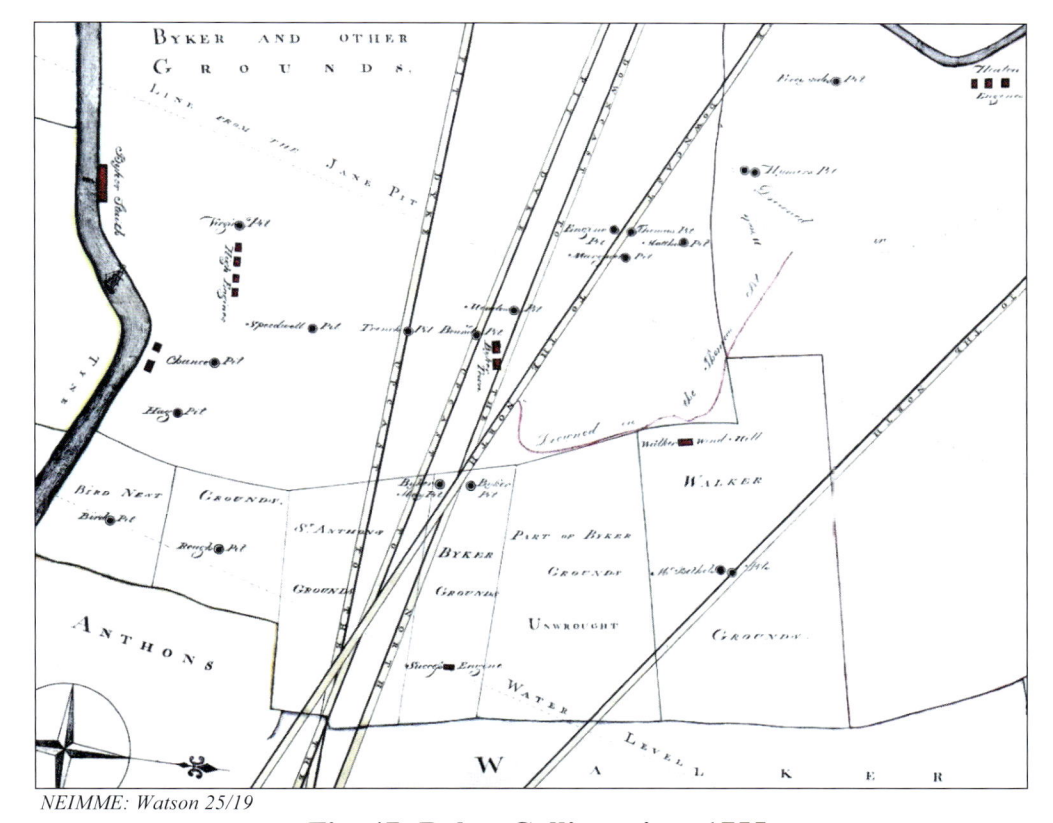

Fig. 47: Byker Colliery circa 1755

---

[168] NEIMME: Watson 2/4/66; East 10b, Watson 2/7/82; Newcastle Courant January 2nd 1768.

91

Byker, where the Ridley family mined on the western rim of the Tyne Basin, had a long tradition of using Newcomen engines: the three visited by Sir John Clerk in April 1724 were situated at Old Byker Colliery which straddled Shields Road (fig.46). In the late 1740's, a major redevelopment took place which resulted in the building of two engines at Dents Hole near the Chance Pit and four engines known as the High Engines further west (fig.47). After the closure of Heaton Banks Colliery in 1745, Matthew Ridley bought the three engines in Low Heaton to the north. By 1749, additional assistance was provided by the three engines erected by William Newton for the Grand Allies at Friar's Goose Colliery, across the river in Gateshead, opposite Byker staith. In 1750, Ridley opened the Success Pit on Walker Hill in the eastern part of Byker which had a double engine house. Ridley's colliery was extended northwards into Sir John Lawson's land marked 'unwrought ground' on fig. 47. The Recovery Pit was sunk in 1753 at Walker Hill and the Delight Pit added by 1758. This colliery became known as Lawson's Main. By the 1760's, William Brown was acting as Matthew Ridley's viewer and Byker Colliery was well established as one of the major enterprises on Tyneside. The first Boulton and Watt engine in the Northumberland and Durham coalfield was erected at Byker Colliery in 1778, almost certainly on the orders of William Brown, to assist with pumping. The location of this engine is a mystery but it is very likely that it was a replacement for one of the many Newcomen engines at this colliery.[169]

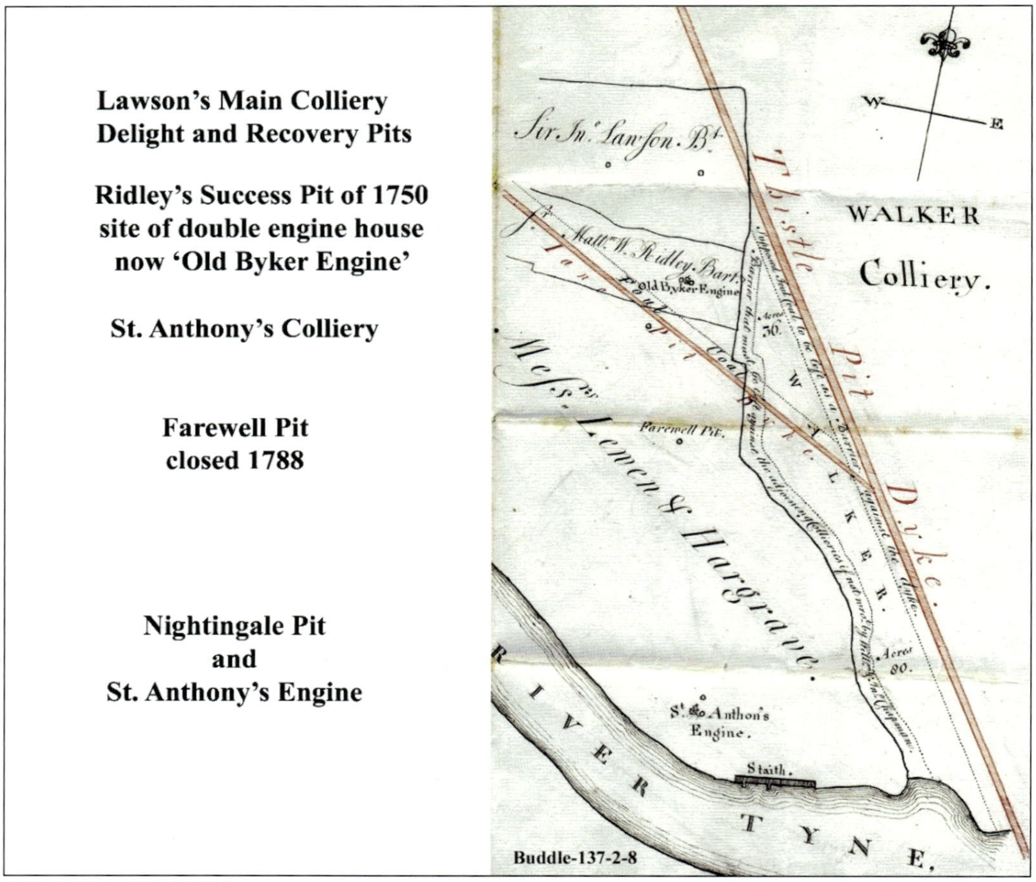
Lawson's Main Colliery
Delight and Recovery Pits
Ridley's Success Pit of 1750
site of double engine house
now 'Old Byker Engine'
St. Anthony's Colliery
Farewell Pit
closed 1788
Nightingale Pit
and
St. Anthony's Engine
W
E
WALKER
Colliery.
Thistle Pit Dyke.
Old Byker Engine
Farewell Pit.
Acres 30.
Acres 90.
St. Anthon's Engine.
Staith.
RIVER TYNE.
Buddle-137-2-8

NEIMME: Buddle 137/2/8

Fig. 48: Byker St. Anthony's Colliery circa 1790

[169] TNS Vol.64 (1992) p.53-75 Jennifer Tann 'The Steam Engine on Tyneside in the Industrial Revolution'. The Lawson family were the principal royalty owners in Byker and confusingly there was more than Lawson's Main Pit.

In 1765, Brown negotiated Ridley's acquisition of St. Anthony's Colliery from the Grand Allies and a winning was made by means of an outstroke from the Success Pit on Walker Hill. Brown commented to Ridley that the coal in St. Anthony's was 'as good as those you have in Walker' which referred to Lawson's Main, his colliery on Walker Hill. Later, St. Anthony's Colliery was operated by the Chapman brothers, William and John, who appointed John Allen as their viewer. Allen may well have been part of Brown's team for he had worked at Hartley Colliery, where he experimented with a stone boiler, at Ford Colliery, and at Low Fell, where Brown had built an engine. At first, the owners of St. Anthony's paid rent for the use of Ridley's engine at the Success Pit but, in the late 1780's, John Allen sank an engine pit near the river. The Nightingale Pit nearby, and later the later Restoration Pit, became the mainstay of the colliery until its closure in 1801. Unfortunately, John Allen was killed in an explosion at St. Anthony's. George Johnson, another of Brown's proteges, replaced him as viewer at both St. Anthony's and Wallsend, the Chapman's other colliery. St. Anthony's Colliery produced about 10,000 chaldrons per annum during the period 1782 and 1801. By this time George Johnson was part owner of St. Anthony's as well as both Bigges Main and Heaton Main collieries; and, as Sam Haggerston remarked, 'he got very rich…and left off going down the pits', which is exactly what his mentor William Brown had done. Unfortunately, the discovery that the workings of St. Anthony's had penetrated beneath the River Tyne into Felling Colliery, belonging to the powerful Brandling family, and a similar trespass had taken place from Heaton Main Colliery into the royalty of the more powerful Grand Allies, led 'to the great grief of Johnson'. He 'was called to London to attend trials on those trespasses after which he retired to Wolverhampton where he died shortly afterwards'.[170]

NEIMME: Watson 20A/20

Fig. 49: Walker Colliery Engines in 1780

Walker Colliery, which lay to the east of Byker and St. Anthony's, had been won by a long drainage level from the Stotts Burn at the beginning of the eighteenth century. In April 1713, Richard Peck provided Hugh Bethell with an estimate for 200 yards of open cut and 1,868 yards of drift to 'the two present pitts now sinking'.[171] These pits were to the rise of the Thistle Pit dyke; but in the eastern part of Walker royalty, where the High Main coal was 100 fathoms deep, the coal was inaccessible at that date. The deeper coal in the eastern part of the royalty was won at the West Engine Pit in January 1762 by the partnership of William Ord, William Peareth and Joseph Reay. William Brown was the engineer chiefly responsible for this development and he was presented with a

---

[170] NEIMME: LBWB II p.8, 21, 42. Brown was also responsible for Ridley's coal mines at Plessey, Tanfield Moor Edge and Collierley. The leading accounts survive for St. Anthony's from 1782 to 1801 ref. TWAMS 3415/CK/2/618. NEIMME: East 3b Dunn 'History of the Viewers' p.32-3 based on the recollection of an old miner, Sam Haggerston.

[171] Peck I p.22

silver punch ladle by the partnership which is now part of the collection of the Society of Antiquaries of Newcastle. By January 1763, the West Engine Pit was in operation while the East Engine Pit and Anne Pit were being sunk (fig.50). In December 1762, a 73 inch cylinder for the second fire engine was shipped from Coalbrookdale foundry to Wincomblee on the River Tyne.[172] This cylinder, which weighed 6 ½ tons, surpassed anything which had been seen in the north and such was its size that three jets of water were needed for condensing the steam. Brown appreciated the importance of an adequate supply of steam to the engine which was achieved through building multiple boilers: four were in use for the Walker engines. When Gabriel Jars, the distinguished mining engineer and member of the French Academy of Sciences, visited Walker in 1765, he commented that the engine was the largest in Europe and noted that three of the boilers were always in fire while the fourth was under repair.[173] The double engine house was 45 feet long, 26 feet wide and 38 feet high; the walls were three feet thick with the exception of the north wall, five feet thick, which housed the beams. In addition, there were two smiths' shops and a leanto 52' long.[174] The engine house is shown on a map of Walker dated 1780 (fig.49).[175]

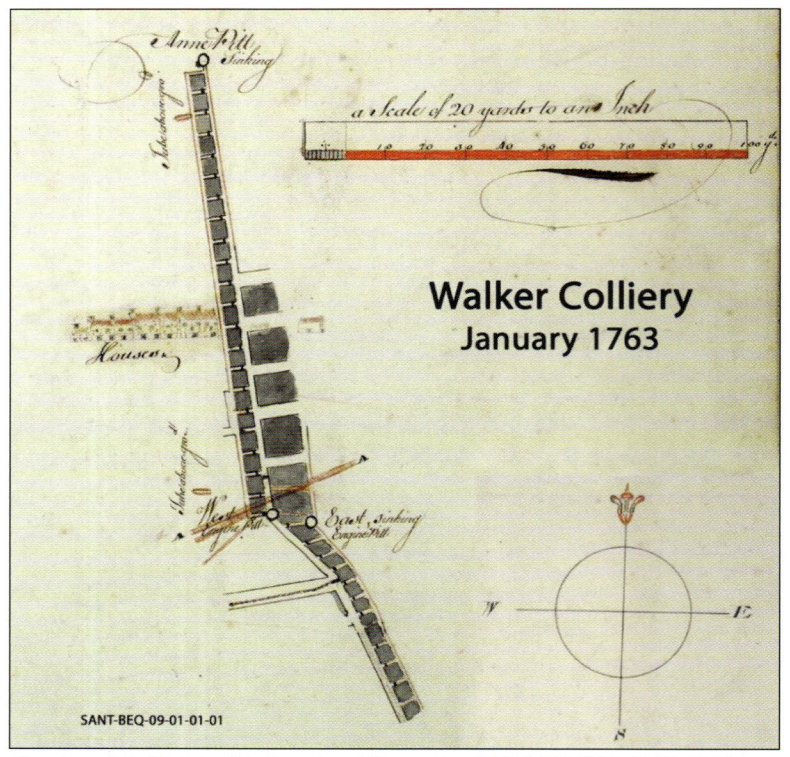

Courtesy of the Society of Antiquaries of Newcastle.: NRO-SANT-BEQ-09-01-01-01
Fig. 50: Walker Colliery in 1763

---

[172] Brown list (fig.54) recorded the size of the cylinders of the two engines at Walker as 72'' and 73'' but Jars and others subsequently have stated that the size of the larger cylinder was 74''. The most obvious route to transport such a large object was down the River Seven to Bristol by barge and then by ship round the south coast to Tyneside. However, Ken Rogers 'The Newcomen Engine in the West of England', which is based on the diary of Thomas Goldney the agent for Coalbrookdale at Bristol, recorded (p.49) that 'the account book refers in a marginal note to the sending of this cylinder… but it does not appear to have passed through Bristol'. Four cylinders were shipped to Isaac Thompson in 1765 via Bristol: a 75'' one which was probably for Benwell, a 60'' one probably for East Denton not Benton, a 64'' one probably for Lambton and a 66'' one probably for Duddingstone in Scotland.

[173] G. Jars 'Voyages Metallurgiques' Vol.I p.195. He recorded that the cylinder was 74'' diameter.

[174] NEIMME: Forster 1/4/180.

[175] NEIMME: Watson 20A/20.

The method of drawing coal from a depth of 100 fathoms also attracted the attention of Gabriel Jars, who noted the uniqueness of the gin, which was driven by eight horses. This engine could wind a corf containing 6 cwt of coal from a depth of 600 feet in two minutes. However, the task placed great strain on the horses and in May 1768, Dr. Rotherams proposed building 'a new machine for drawing coals at Walker Colliery…with the assistance of my friend Mr. Brown'. He pointed out that his invention would take all the weight of the rope and reduce the number of horses needed.[176] Subsequently this machine was replaced by a water gin, the most common form of haulage machinery until the development of rotary motion by James Watt.

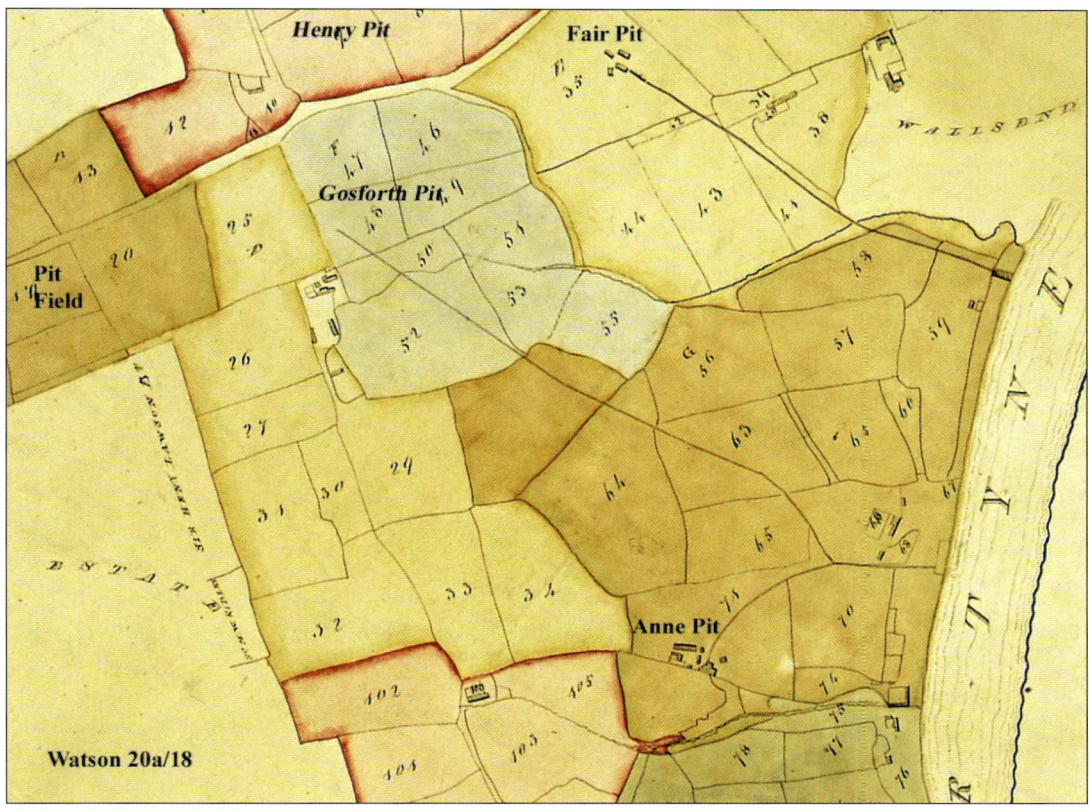

NEIMME: Watson 20A/18

Fig. 51: Brown's Development of Walker Estate circa 1780

Walker Colliery continued to develop under Brown's direction. The Fair Pit was probably sunk in the early 1770's and a waggonway constructed to a new staith at the north east corner of the estate. The Gosforth Pit was won in April 1780 and a waggonway built to the old staith at Wincomblee. The colliery's share of the vend increased from 21,000 chaldrons in 1773 to 35,000 by 1804 making Walker Colliery, like the neighbouring collieries associated with the Brown family at Willington, Wallsend and Bigges Main, one of the largest coalmines in the world. By 1796, there were two Boulton and Watt engines at work at the King Pit in the southern part of the estate, one for pumping and the other for winding.

In 1773, William Brown and Matthew Bell, in partnership with the Newcastle attorney William Gibson, leased the coal under Sir Ralph Milbank's estate in Willington for 31

---

[176] NEIMME: Forster 1/19/E6. TNS Vol.64 (1992) p.56 Jennifer Tann 'The Steam Engine on Tyneside in the Industrial Revolution'.

years. The consortium also leased Willington farm and Backwell farm for 30 years from May 1774 to produce feed for the many horses needed for both the colliery and the waggonway. George Johnson of Byker, Brown's viewer at Throckley Colliery, also served as viewer at Willington. A document, dated September 1771, in the papers of John Watson, gives an estimate for the winning and working of Willington Colliery. It is interesting not only because it illustrates the costs involved in establishing a major colliery during the late eighteenth century, but also because it shows that the main components, such as the pumping engine, the waggonway, the waggons and even the staith, were items recycled from elsewhere.

Estimate for Winning and Working Sir Ralph Milbank's Colliery near Willington in 1771

Boring	200
Offtake Drift	1000
Engine pit with Timber	1200
Engine Compleat	2500
Two Gins and two pair of Ropes	200
Stock of Materials for two Pitts	100
Horses for two Pitts say 40 at 10 £ each	400
1 ¼ Mile of Way main and By Way	800
Staith and Key	800
15 Waggons	150
Contingency unforeseen	650
	8000

Stock in Hand

Engine	2000
Gins	150
Materials	100
Horses	400
Waggonways	800
Staith	300
Waggons	150
Materials to Spare	100
	4000
	£4000 to expend

The principal source of the pumping engines was Throckley Colliery which was in the process of a major reorganisation which included the dismantling of the Delight or East Engine and the West Engine. Entries in George Johnson's view book indicate that boilers, pumps, pipes and a cylinder were transported by keel boat from Lemington to Willington.[177]

---

[177] NEIMME: George Johnson's viewbook.
1st April 1775
The Delight Engine boiler is got out and will be upon the Carriage and down to Newburn if possible this day and as the tides will be good on Monday and Tuesday it will be a proper time to get her to down to Willington.
29th April 1775
The West Engine boiler is got down to the East Engine to be repaired for Willington.
20th May 1775
The west Engine Cylinder, several pumps and small pipes were put into a Keel yesterday and will be at Willington this day; the boiler will be at Newburn and put into a Keel this day.

On 23rd August 1773, William Gibson, one of William Brown's assistants, noted in his diary that he had 'designed branch of way from the Engine pit to join the partnershipway' at Willington Square and added amusingly that on the following Monday 'alterations were made by Mr Brown'. The Partnership Way was the Grand Allies line from East Longbenton Colliery to Willington Quay built in 1762. Willington Colliery did not open until November 1775; and what this note shows, apart from the fact that clearly the assistant had to be kept in order, is that the waggonway was planned at an early stage in the development of the colliery. This was necessary because the railway was needed to bring building materials to where the shaft was being sunk, the colliery buildings erected and the mining community established.[178] Sam Haggerston recorded that in 1775 William Brown 'built two engines at Willington and won that Colliery' – an event which was marked by the customary celebratory dinner for the workmen. On 2nd November 1775 the owners donated 'a fat ox roasted, a large quantity of ale and a waggon load of punch' to celebrate the winning of the Engine Pit. Gibson's comments were similar to Mrs Montagu's account of the winning of East Denton:

> 'At noon the people all marched in formation to Mr. Brown's and got each a gill of punch by way of drinking success to Willington Colliery. Afterwards returned to the Granaries and swimmed in Ale till eight o'clock.'[179]

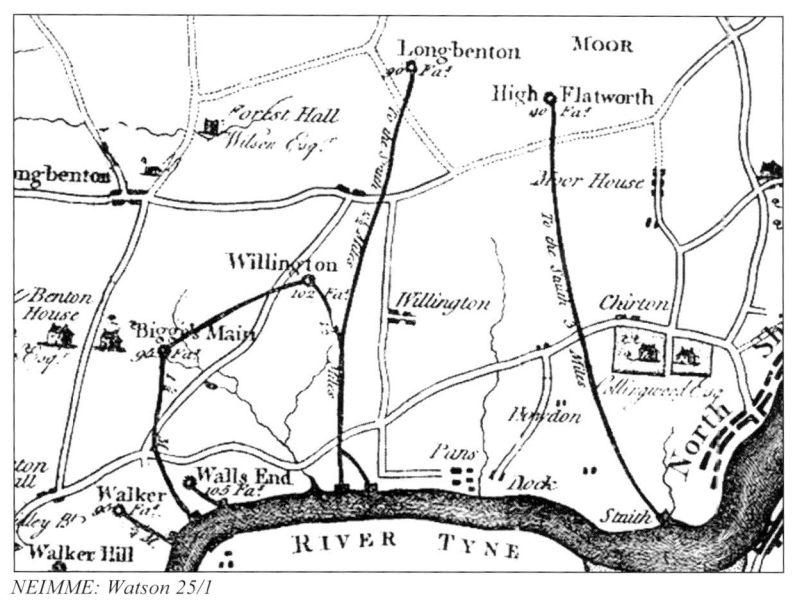

NEIMME: Watson 25/1

Fig. 52: John Gibson's Plan showing Willington in 1787

During the next decade, Willington Colliery expanded westwards with the sinking of the Edward Pit and the Bigge Pit; and another branch was added to the Partnership Way from the west. After William Brown's death, the westward expansion of Willington Colliery was continued into East Benton, Little Benton and Longbenton by his son. Shafts for a new colliery, known as Bigges Main, were sunk between 1784 and 1786. The waggonway was built southwards through Wallsend to Carville Shore. This development created an opportunity to provide Willington Colliery with an independent

---

[178] NEIMME: William Gibson Diaries. Note this man was the viewer at Willington and should not be confused with William Gibson, the attorney and Town Clerk of Newcastle upon Tyne, who was one of the owners. The partnership referred to was the Grand Allies.

[179] NEIMME: Gibson Diaries.

route to the River Tyne: a branch was built westwards from the Bigge Pit to join the Bigges Main Waggonway at the colliery village (fig.52). The alternative route also releaved the pressure of traffic on the section of the Benton Way serving both collieries.

William Brown also provided engines for collieries on the south side of the river trying to win coal from Tyne Basin. In 1775, he built a double enginehouse at Washington, where his son Richard was part owner, but only one engine was needed. In the following year, he supplied an engine for Felling Colliery owned by the Brandling family. William Brown died in February 1782 a few months before Wallsend Colliery was won. However, with the assistance of William Gibson, he had played a part in the development of this important colliery which would soon become synonymous with the best quality household coal on the market.

The conquest of the Tyne Basin was achieved through William Brown's mastery of the Newcomen engine. Old Sam Haggerston listed 21 collieries in the Great Northern Coalfield and three in Scotland where William Brown had built fire engines which made him one of the principal exponents of this important art during mid eighteenth century. This new technology, together with the parallel development of waggonways, made possible the great expansion of the coal trade in Northumberland and Durham and transformed South East Northumberland into the major coal shipping point for the Great Northern Coalfield. From the fragmentary evidence which survives, it is clear that William Brown of Throckley was one of the most accomplished engineers of his generation, whose expertise was valued far beyond the region. He was a friend of some of the icons of the Industrial Revolution, great engineers of the calibre of James Brindley, John Smeaton and Abraham Darby; and like them he deserves not only national but international recognition.

NEIMME: Hair

Fig. 53: Wallsend 'A' Pit: Probably William Brown's Last Major Job

Appendix I: The Newcomen Engines in the Great Northern Coalfield

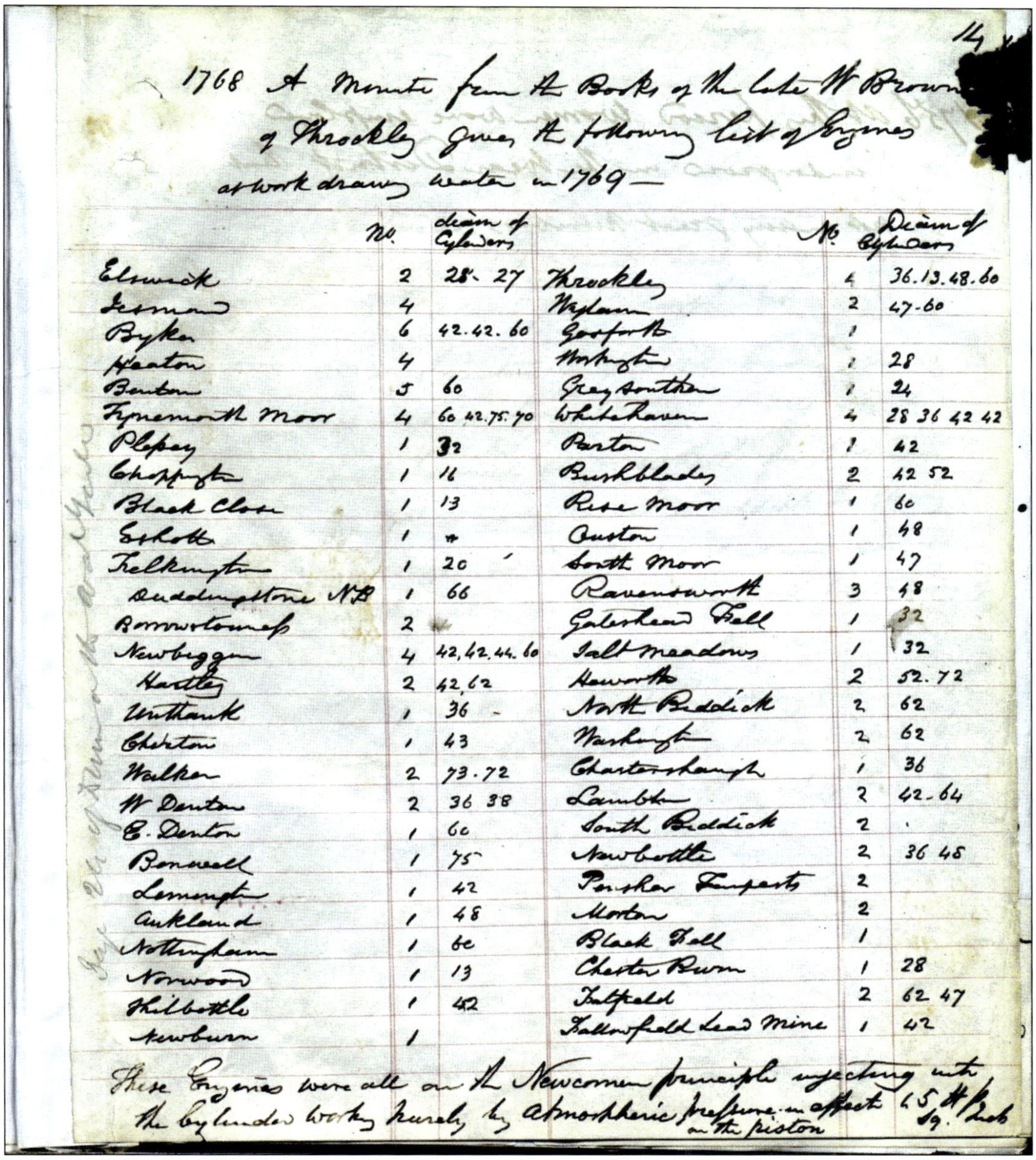

NEIMME: East 8B

Fig. 54: Brown's List of Newcomen Engines

One of the principal sources for the location of Newcomen fire engines in the Great Northern Coalfield, and William Brown's role in their installation, is Matthias Dunn's manuscript 'The History of the Viewers'. The information comes in two forms: the recollections of Sam Haggerston and a minute from Brown's notebooks, Brown's list. This manuscript was subsequently used in the nineteenth century by Dunn and Galloway in their histories of the coal trade, and later by Raistrick, Mott, Allen and Tann in their papers to the Newcomen Society.[180] Consequently, it has had a

---

[180] Dunn M. 'View of the Coal Trade', Newcastle 1844; 'Treatise on the Winning and Working of Collieries' Newcastle 1848; Galloway R., 'Annals of Coalmining', London 1898; Trans. Newcomen Society 17: Raistrick A. R. 'The Steam Engine on Tyneside: 1715-1778; T.N.S. 35: Mott, R.A. 'The Newcomen Engine in the Eighteenth Century'; T.N.S. 42, 43, 45: Allen, J.S. 'The Introduction of the Newcomen Engine in the Eighteenth Century; T.N.S.: Tann J. 'The Steam Engine on Tyneside in the Industrial Revolution'.

considerable influence. However, without local knowledge, it is not easy to fully understand this document and mistakes have been made. This point is well illustrated in the maps provided by both Raistrick and Mott which are far from accurate. Furthermore, Raistrick's paper claimed to provide a 'chronological list of engines in the Northumberland and Durham Coalfield from 1715 to 1778'; and this was based upon Dunn's manuscript and the reports of the viewers in the Mining Institute at Newcastle. It is irritating that he provided no references for his listings; but, that apart, there are major errors in his work. His geographical knowledge is often inaccurate (Nutt Hall is in Nottinghamshire not the North East); the engine at Elsdon in rural Northumberland was a bob gin not a fire engine; and it is simply not true that a large number of collieries on Brown's list are not otherwise known for maps exist for most of them as the table below illustrates. During the course of this research, an examination of the extensive map collections once belonging to William Brown and John Watson has been undertaken to locate the sites of the engines on Brown's list and thereby verify their existence. Also, a review of the note books of the viewers was carried out to seek additional verification. An examination of these primary sources confirms the accuracy of Haggerston's memory and the credibility of Brown's list. The research also raises questions about the interpretations which have been made by subsequent authors using the manuscript; and in particular their views on the number of Newcomen engines at work in the Great Northern Coalfield during Brown's time.[181]

Sam Haggerston, the old furnace keeper at Hebburn Colliery, where Matthias Dunn was underviewer in 1811, commented that, in about 1750, William Brown 'began to turn his attention very much to the improvement and application of the Steam Engine which was then in a very rude state'. Sam recorded that Nicholas Walton, the agent of the Grand Allies, built a double engine at Allerdean (Ravensworth) and three engines at Friar's Goose in about 1750; and then gives an account of the twenty nine engines built by Brown between 1756 and 1776.[182] Because there was often a considerable time gap between planning the engine, sinking the shaft, driving the drainage level, building reservoirs and water courses, erecting the engine house and installing the engine, the precise dates given by Haggerston need to be treated with caution. For example, he cites 1758 as the date for building the Walker engine but, although the engine house was planned by that time, the first of the cylinders did not arrive from Coalbrookdale until 1762 and a map dated January 1763 (fig.50) shows that, although the west engine was built, the east engine was unfinished. All of the engines named by Haggerston are on Brown's list but confusingly some are given a different name which is why local knowledge is needed to interpret this document: Birtley North is Black Fell; Bells Close is Denton; Walbottle is Newburn; Shiremoor is Tynemouth Moor; Beamish is South Moor; and Low Fell is Gateshead Fell. The primary evidence suggests that Sam Haggeston's recollections were accurate: the existence of all of the twenty nine engines is confirmed either by the viewer's reports or their maps; and no additional engines have been found in the primary sources which are attributable to Brown. Posterity

---

[181] NEIMME: East 3b: Dunn M, 'The History of the Viewers'; Dunn M., 'View of the Coal Trade', Newcastle 1848; Galloway R., 'Annals of Coal Mining and the Coal Trade', London 1898; Transactions of the Newcomen Society, Vol. 17 Raistrick, Vol. 35 Mott, Vols. 42,3,5 Allen, Vol. 64 Tann. NEIMME: Watson volumes 19 to 34; Forster 1/4 and 1/5, Buddle 13,15 and 25, Peck. NCRO: SANT/BEQ/9/1/1 and 2 and 3.

[182] NEIMME: East 3b: Dunn M, 'The History of the Viewers'; Dunn M. The engines built by Brown are given as Throckley in 1756; Birtley North (2), Old Lambton and Byker in 1757; Walker (2) and Bells Close in 1758; Heworth (2) in 1759, Shiremoor (2) and Hartley in 1760; Walbottle (2), Oxclose, Beamish and Benwell in 1762; West Auckland in 1763; North Biddick, Pittenweem, Bowness, Musselburgh and Low Fell in 1764; Lambton in 1766; Fatfield in 1772; Washington and Willington (2) in 1775; and Felling in 1776. Note that 26 not 22 of these engines were in the Great Northern Coalfield.

should be grateful to old Sam who sadly died the following year in the workhouse at Heworth.

The second element in Dunn's 'History of the Viewers', is entitled 'A Minute from the Books of the late W. Brown' (fig.54). This page consists of three parts: a title, a list of collieries with a note of the number of engines built and the size of their cylinders, and a concluding statement that the 'Engines were all on the Newcomen principle'. The list is the work of Brown, the title and end piece were written by Dunn who was without doubt a very eminent engineer but he was a much less competent historian. The title is curious for several reasons and provides some evidence for this assessment of Dunn the historian. It declares that what follows is a 'list of Engines at work drawing water in 1769' but it appears to be dated 1768. The list includes most of the early engines on Tyneside but these are unlikely to have been still working in 1769, given that it cost about £400 p.a. to keep an engine in running order. Elswick Colliery had been drowned before 1740; Jesmond and Heaton Banks collieries lost their battles with water in 1745, although two of the Heaton engines were probably still working in 1761 to assist Ridley's colliery in Byker; by 1769 Humble's engine house at Lemington had been converted into tenements; West Longbenton Colliery was replaced by East Longbenton in 1762 and the engines sold. Galloway, who claimed that the list was compiled for John Smeaton, recognised that 'many of the engines in the list had been worn out and given up and that the number actually at work was 57'. So Dunn's historical comments have to be treated with caution. It is my belief that Brown was not recording the engines at work in 1769 but the collieries where engines were working, or had been working, since the invention of the Newcomen engine.

Dunn's title is far from accurate but, in contrast, Brown's list is very sound: this is the record of an acknowledged expert on the subject of fire engines who, through his work as a viewer, was intimately associated with all the major collieries in Northumberland and Durham. The list appears to be a comprehensive statement of the engines known to have existed in the Great Northern Coalfield during his lifetime, plus engines from elsewhere, which were known to him through his work further afield. Eighty of the ninety nine engines listed are at collieries in the Great Northern Coalfield. All of the collieries in that coalfield which are known from other primary sources to have used Newcomen engines by 1769 are on Brown's list, except Gateshead Park. This is a curious omission for Brown was certainly aware of the of the existence of fire engines at this colliery: in a letter to Spedding in 1755, he commented that the Park had been laid in costing the Grand Allies £20,000 despite there being 'three fire engines and these of large dimension upon her'.[183] Doubtless, Brown is referring here to the engines built by Nicholas Walton in 1749. He does record an engine at Saltmeadows which was once part of Gateshead Park. Also, some of the engines at collieries on the list built fifty years earlier, before Brown embarked upon his career as an engine builder, appear to be absent from the list: these are Byker's first three engines and Beighton's Oxclose engine. There is a group of engines in Cumbria where Brown had a long association through his friendship with the Spedding family: Workington, Grey Southern, Whitehaven and Parton. There is another interesting group of engines with small cylinders serving landsale collieries in central and north Northumberland, where Brown is known to have been active: at Plessey, Choppington, Black Close, Eshott, Felkington, Unthank and Shilbottle. There are three engines in Scotland near Edinburgh where

---

[183] NEIMME: WBLB I p. 262 William Brown to Carlisle Spedding 10th April 1755.

Brown had long standing business connections. Although Brown made several visits to Fifeshire, he does not name an engine at Pittenweem Colliery, which was included amongst the three Scottish engines named by Haggerston. The other engines were at Nutt Hall in Nottingshire, where one of Brown's viewers was manager, and Fallowfield lead mine in the North Tyne valley. Curiously, there is no mention of engines in the Swaledale mineral mines where Brown's correspondence shows he had business interests. Perhaps, they were never built despite Brown's efforts.

The table below records the fifty four collieries on Brown's list. Collieries known to have had engines built before 1752, the date when Brown is thought to have begun building engines, are marked by an asterisk. Engines built between 1769, the date of the list, and Brown's death in 1782 have been added in italics. The ones where Brown is said by Haggerston to have built an engine are highlighted in bold type. Brown is known to have been associated with many of the other collieries, such as Black Close, Wylam, Lemington and West Denton, but it is not known whether he actually built their engines. The dominance of William Brown as a contractor for steam engines in the Great Northern Coalfield between 1752 and 1782 is self evident: he is known to have supplied engines to 17 of the 39 collieries named on the list. The map references enable the location of these engines to be pinpointed and such is the detail of these plans that often an illustration of the engine is included.

Brown's List

North bank of River Tyne		Cylinder Size	Map Reference	Notes
01 Wylam	2	(47, 60)		Fig.18
02 Throckley	**4**	**(36, 13, 48, 60)**	Wat. 27/21	Fifth engine built in 1785 fig.13,14,23
03 Newburn	**1**		Br. 1/24, 29, 33	Walbottle engine fig.17
04 Lemington*	1	(42)	Br. 1/6	Humble's engine fig.7
05 West Denton*	2	(36, 38)	Br. 1/20	John Blackett's engines.
06 East Denton	**1**	**(60)**	Wat. 23A/15; Br. 1/22	Later 2nd engine north of Main Dyke
07 Newbiggin	4	(42, 42, 44, 60)	Wat. 23A/12 and 14	Some bought from Benton West fig.5
08 Benwell	**1**	**(75)**	Bell 14/135	Marked old engine in west.
09 Elswick*	2	(25, 27)	Wat. 28/5	Allen: erected 1718; Wortley's engines
10 Jesmond*	4		NCRO: ZRI 50/9	Allen: two erected by 1731 fig.39
11 Heaton Banks*	4		Wat. 25/13	Built 1729-3; fig.39,41
12 Byker*	**6**	**(42, 42, 60)**	Wat. 25/19, 27/24	A complicated site see notes fig.46,47
13 South Gosforth*	1		Wat. 20A/9	Drained Benton West fig.42
14a Benton West*	3		Wat. 20A/9, 13	Third engine added by 1762 fig.37,42
14b Benton East	2	(60)	Wat. 19/16, 20A/2	Replaced Benton West in 1762
15 Walker	**2**	**(73, 72)**	Br. 1/1 and 57	See text p. fig.49,50,51
16 Shiremoor	**4**	**(60, 42, 75, 70)**	Wat. 21/1	See text p. fig.45
17 Hartley	**2**	42, **62**	Br. 1/69; Wat.19/11,14	Fig.43. Three engines on Wat.19/11
18 Chirton	1	(43)	Wat. 21/6	Brown's pits marked in Preston
Callerton	2		Br. 1/24; Bell 14/497	Extension of Throckley Moor Colliery
Byker St. Anthony's	1		Buddle 137-2-8	Fig.48
Willington	1		Br. 1/62	Completed 1775
Wallsend	1		Wat. 22/1	Completed 1783
Bigges Main	1		Wat. 22/28	Completed 1786
South Bank of River Tyne				
19 Bushblades*	2	(42, 52)	Br. 3/75; Wat. 31/10, 34/25	Complicated – see notes Fig 38,56,57
20 Risemoor	1	(60)	Br. 3/93 and 94	Weighton/Walters engine after 1763
21 Ravenswood	3	(48)	TWAMS: DT.Bell/2/183	Allerdean engines
22 Norwood*	1	(13)		Allen: erected 1718
23 Gateshead Fell	**1**	**(32)**	Br. 3/10, Bell 19/113	Low Fell near Ann Pit
24 Saltmeadows	1	(32)	Br. 3/1, 3/15	Claxton's engine
25 Heworth	**2**	**(52, 72)**	Br. 3/15, Wat. 30/7	Known as Jarrow Colliery
26 Washington*	**1**		(62)	Br. 3/19, Buddle 7/1	Successor to Oxclose engine
*Friar's Goose**	3		Br. 3/15, Wat. 29/1	Operated 1747 - 1763
Felling	1			Dunn p.21 – sunk 1776-9

North Bank of River Wear

27 **South Moor**	1	(47)		Dunn p.20 – Beamish Colliery
28 Chesterburn*	1	(28)		Flatts erected 1723
29 Ouston	1	(48)	Wat. 31	Pelton Moor
30 **Black Fell***	2		Br. 1/14, Wat. 31/3	Toft Hill Col. North Birtley, Dunn p.19
31 **North Biddick***	2	(62)	B/T 1737, Wat.29/35, Br.3/50	Hedworth's engine, fig. 40
32 Chartershaugh*	1	(36)	B/T 1737, Br.3/50	Peareth's engine, fig. 40
33 **Fatfield**	2	(62, 47)	Br. 3/17	Dunn p.20, Old engine pit 1767
Rickleton	1		Br. 1/17	

South Bank of River Wear

34 **Lambton***	2	(42, 64)	B/T 1737; Br. 3/50	Esq. Lambton's engine. Dunn p.20
35 South Biddick*	2		B/T 1737 Br. 3/50	Nicholas Lambton's engine, fig. 40
36 Newbottle*	2	(36, 48)	Br. 3/52	Allen: erected by 1733
37 Norton *	2		Br. 3/52	Usually known as Morton or Lumley
38 Pensher Tempest	2			
39 **West Auckland**	1	48	Br. 2/39	Dunn p.20

Engines in other Coalfields

40 **Duddingston**	1	66		Dunn p.20 lists these Scottish engines
41 **Bowness**	2			as Pittenweem, Bowness, Musselburgh
42 **Unthank**	1	36	Br. 2/15; Wat.24/18	Boghill Colliery 1766
43 Felkington	1	20		
44 Shilbottle	1	42	Br. 1/16	1764 no engine - new winning planned
45 Eshott	1			
46 Choppington	1	16		
47 Black Close	1	13	Wat. 23/10, 24/36	Built between after 1755. Fig.33
48 Fallowfield	1	42		For a lead mine
49 Plessey	1	32	Br. 1/8	Ref. to engine but not marked 1763
50 Workington	1	28		
51 Gray Suthern	1	24	Br. 2/31	
52 Whitehaven	4	28, 36, 42, 42		
53 Parton	1	42		
54 Nottingham	1	60	Br. 2/57 – 60	Nutt Hall Colliery

Allen T.S. Transactions of the Newcomen Society Volume 42 p.169 – 190
B/T Burleigh and Thompson's Plan of the River Wear 1737
Br. Northumberland County Record Office – Brown's Plans – Sant/Beq/ 9/1/1 and 2 and 3
Bell NEIMME – Bell papers
Buddle NEIMME – Buddle papers
Dunn NEIMME – History of the Viewers – East 3b
Wat. NEIMME – Watson papers

The number of Newcomen engines working in the Great Northern Coalfield has been the subject of much debate. Any assessment is complicated by two factors: the replacement of engines either worn out or no longer adequate; and the movement of engines within a colliery and from one colliery to another. Some engines were certainly built as replacements: the Bushblades engine of 1730 belongs to this category. Engines were moved after they had fulfilled their purpose or were found to be unsuccessful where they were first installed. Brown recorded in a report on Risemoor Colliery that Humble and Hodgson 'erected a Fire Engine on the Spring Pitt in that colliery which did not answer their Expectations and at last they desisted from working'. It was probably moved to one of Humble's other collieries (Lemington, Shiremoor or North Birtley) or sold. Certainly, others were moved when no longer needed at a particular site: parts of the Throckley engines were moved to Willington Colliery by Brown, parts of the West Longbenton engines were sold to Holywell Colliery and the Houghton engine was moved north nearer Newbottle. This transfer of machinery needs to be borne in mind when trying to assess the number of engines built for the industry. It may be more accurate to list collieries using engines as Brown did. The history of the coalfield in the eighteenth century, supported by the evidence of the maps, the viewers' reports and Brown's list, would suggest that the number of engines working within the coalfield at that time has been exaggerated. It has been claimed that 26 engines were in operation in the coalfield between 1712 and 1733: the figure is more likely to have been 13.[184] It is also claimed that 133 engines were erected between 1734 and 1775. However, the number of seasale collieries, not all using fire engines, ranged from about 16 to 36 during this period and therefore the figure is likely to be nearer half that number. For example, an examination of the collieries supplying the vend from the River Tyne in 1773 and the River Wear in 1772 shows that there were only 36 collieries involved in the seasale trade and that there were probably only 38 engines at work. Even allowing for the fact these engines wore out and were replaced, the number of engines built during this period is not likely to have been 133.[185]

Additional Notes on Brown's List

Byker This was a very wet colliery which relied heavily on pumping engines from the time Newcomen's engine first became available. In 1717, the Swedish engineer Martin Trievald built the first engine at Byker.[186] In April 1724, three engines are recorded by Sir John Clerk[187] one of which was undoubtedly Trievald's engine. These three engine were in Old Byker Colliery on the either side of Shields Road (fig.46) and the Kenton, Thomas and Jane pits are probable locations. They do not appear to be on Brown's list.

In July 1746, an 'estimate for winning Byker Colliery in Ridley's Byker further east without coming through St. Anthonys ground' was £4,150, which included the cost of two engines at the Success Pit on Walker Hill. However, 'if Byker Colliery be laid in and three fire Engines more erected' and it would cost an additional £3,600 for these

---

[184] The early engines were at Tanfield Lea, Oxclose alias Washington, Byker (3), Elswick (2), Norwood, Flatts (2), Gateshead Park (1 or 2), Heaton (4), Houghton alias Newbottle.
[185] Flynn M.W. The History of the British Coal Industry Vol.2 (Oxford 1984) p.119-28 and in particular the references to the work of Dr.J.W. Kanefsky.

Quantities Stipulated in Chaldrons for the Vend of the River Tyne in 1773

1. Grand Allies: Longbenton; Tanfield	65,000	(4)	2. Lady Windsor: Pontop	32,000	
3. Mr Simpson: Tanfield Moor Edge	16,000	(1)	4. Bushblades	14,000	(2)
5. Whitefield: Chopwell	18,000		6. Lord Kiery: Tanfield Townhead	13,000	
7. Duke of Northumberland: Walbottle	16,000	(2)	8. Greenwich Moor: Throckley	17,000	(2)
9. Preston Moor	17,000		10. Cramlington: Hollywell	13,000	(4)
11. Wylam	13,000	(2)	12. Montagu: East Denton	16,000	(1)
13. Mr Ridley: Byker	16,000	(1)	14. Walker	21,000	(2)
15. Mr Ormston & Co: Chirton	15,000	(1)	16. Sir Tho. Clavering: Andrews House	6,000	
17. Mr Waters: Risemoor	12,000	(1)	18. Mr Lybourn & Co.	5,000	
19. Lord Ravensworth: Team	26,000	(3)	20. Lord Strathmore: Northbanks	8,000	
21. Mr Pitt: Tanfield Moor	17,000	(1)	22. Chirton	10,000	(1)

Total 386,000 chaldrons (1,022,900 tons) **Collieries 22** **Engines 28**

Coals Vended from the Respective Collieries on the River Wear in 1772

1. Ravensworth & Ptns: Mount Moor	20,655		2. Sir Ralph Milbank: Harraton Moor	22,067	
3. William Lambton Esq: Wylam Moor	21,582	(1)	4. Nicholas Lambton Esq: Biddick	17,699	(1)
5. John Tempest Esq : Rainton	24,826		6. Mrs Jennison & Ptns: Flatts	27,655	
7. William Peareth Esq					
(a) Chatershaugh	13,339	(1)			
(b) Toftmoor (S. Birtley)	18,173		8. Jennison Shafto & Ptns: Washington	24,716	
9. Humble & Smithson: Blackfel l	11,360	(1)	10. Mr Nesham: Houghtonburn Moor	9,371	(1)
11. Shafto & Hudson: North Biddick	143	(2)	12. Morton Davison: South Moor	25,480	(1)
13. John Tempest Esq: Pensher	9,960	(1)	14. Humble & Stafford: Toft Moor	10,247	(1)

Grand Total 257,273 chaldrons (681,773 tons) **Collieries 14** **Engines 10**

---

[186] Newcomen Society Publication.
[187] National Archive of Scotland GD18/2106.

engines and £1,800 for the three shafts.[188] This referred to stopping the engines at Old Byker and/or Dent's Hole.

The six engines on Brown's list are probably at the Thomas Pit, Success Pit, Dent's Hole or Hope Engines (2), and the High Engines (2): later another two were added at the High Engine site (fig.47).

Elswick Wortley-Montagu engines of 1718 were in operation by 1724 near the waterfront. Their exact location is uncertain but it may have been where the gin pit shafts are marked on figure 55. Perhaps the two buildings are converted engine houses? By 1740 the colliery was drowned and there was an estimate for replacing these engines and repairing the Tyne level [189] but there is no evidence that the work was undertaken. The colliery operated on a small scale until it Buddle became manager in 1804.

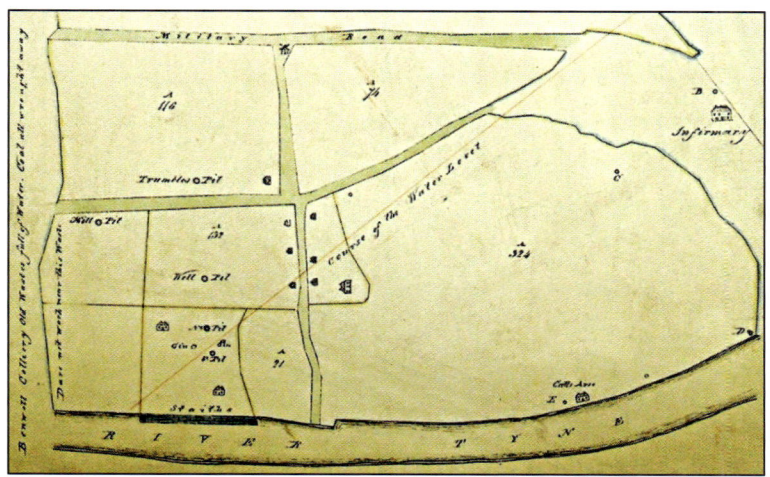

NEIMME: Watson 28/5

Fig. 55: Elswick Colliery circa 1755

Tanfield Lea The first engine is recorded on the map of Gilbert Spearman's estate at Westerleigh in October 1715: Wat. 34/25 (fig.38). Because this is the only evidence of the first engine on Tyneside, some have questioned the authenticity of the map.

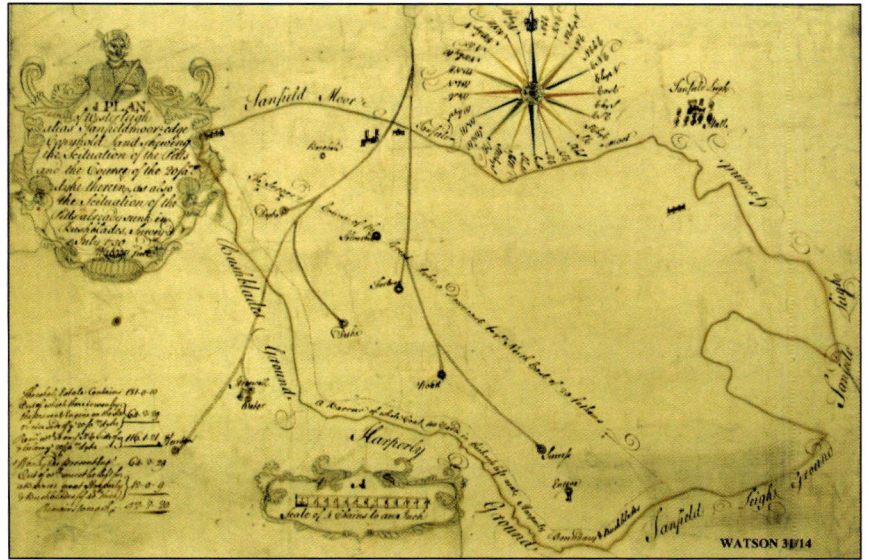

NEIMME: Watson 31/14

Fig. 56: Bushblades Engine of 1730

---
[188] NEIMME: Forster 1/4/113.
[189] NEIMME: Forster1/4/56.

However, there was certainly an engine marked on the same site on John Watson's map [190] dated July 1750 (fig.56). The Bushblades estimate of 1730 reveals that the engine cost £800 and a level 1,200 yards long cost £1,080. [191] There is no price for engine house which would suggest a replacement for the 1715 engine. This drainage level would reach Bushblades to the rise from Westerleigh. By 1729, Richard Ridley had purchased the estate and the entry in the Coalbrookdale stock book referring to Ridley was doubtless for the Bushblades engine.[192] It is possible that Ridley, who had mining interests in the area in association with Lady Clavering, was responsible for the first engine especially since he was well known as an early enthusiast for Newcomen engines.

The map of Bushblades dated 1770 shows Silvertop's engine of 1765 on a different site further west in Bushblades estate (fig.57).[193] These are probably the two Bushblades engines recorded in Brown's list.

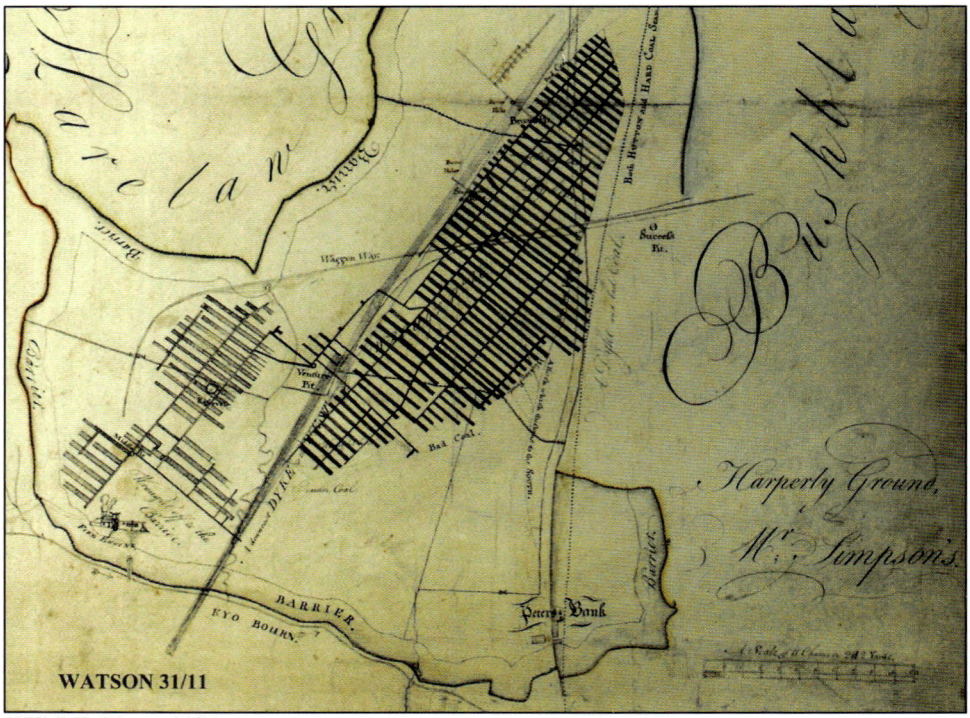

NEIMME: Watson 31/11

Fig. 57: Bushblades Colliery in 1770

Gateshead Park This area between the Newcastle and Durham turnpike road and the stream forming the boundary with Felling estate was the former hunting park of the Bishops of Durham. Two sections, Friar's Goose and Claxtons, were given to St. Edmund's hospital and a third section, known as Saltmeadows, which included most of the valuable river frontage, was leased to the Corporation of Newcastle. From the beginning of Cotesworth's ownership in 1716, the drainage of the colliery was part of a grand scheme to win Heaton in partnership with Henry Liddell who attempted but failed to secure engines, although an engine house had been built by 1718. Following a law suit in 1725, the syndicate agreed to surrender their interest in the colliery, which in the opinion of Dr. Hughes appears not to have been won until a generation later. [194] However, Park was included in the vend of 1725 which would suggest that an engine had arrived. The Minute Book of the Grand Allies records a view by Walton, Barnes, Smith and Claughton in April 1737 which presented a case for a re-opening of the

---

[190] NEIMME: Watson 31/14.
[191] NEIMME: Forster 1/5/39.
[192] T.N.S. 35: Mott Table I No.20.
[193] NEIMME: Watson 31/10.
[194] Archaeologia Aeliana 4th Series 27 (1949) E. Hughes 'The first steam engines in the Durham coalfield'; TWAMS: CK/4/43.

colliery. They noted that the Tyne level 'seems to be standing in most places but both the Engine Shafts are run together' and 'the present Engine house will be of no use'. Clearly, this was a double house but whether it had ever been fitted with engines is not known. The existence of an old engine on the 1753 plan would suggest that it had been (fig. 58). The viewers were uncertain whether one, two or possibly three new engines were needed to win Gateshead Park and Saltmeadows. In 1738, Walton provided an estimate for two engines. Hudderston informs us that Nicholas Walton built three engines in the Friar's Goose in about 1750. Watson's map of Friar's Goose Colliery in 1753 marks the sites of the 'Old Engine' (was this the original engine house?) and the 'New Engine' which had been built by 1749 for Amos Barnes writes an account of the working of the engine. Brown records that the colliery was laid in by April 1755.[195]

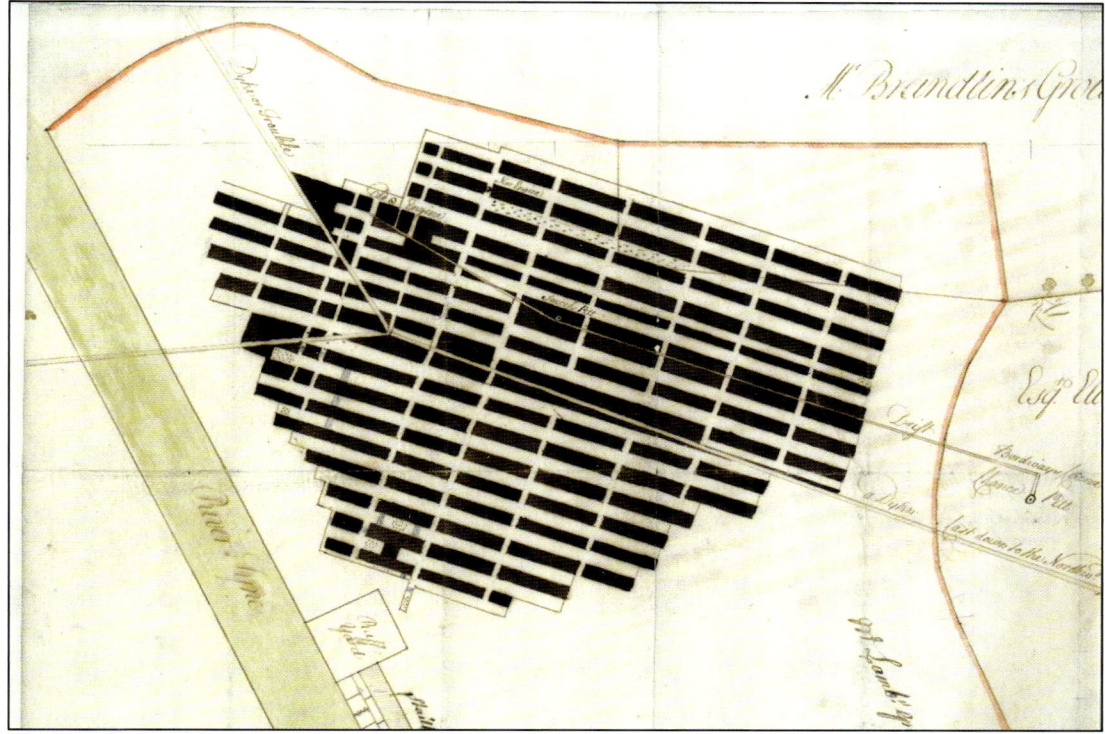

NEIMME: Watson 29/4

Fig. 58: Friar's Goose Colliery 1753

Walton's three engines were probably part of a scheme, which also included the engines at Heaton, to releave the pressure on West Longbenton Colliery which had replaced Heaton Banks Colliery in 1745 as the main source of coal for the Grand Allies in this area. The Friar's Goose engines ceased working in 1763 which co-incided with the Grand Allies move from West to East Longbenton. The location of the three engine pits is marked on a latter plan of 1785. A map in Brown's collection, probably dated to the 1770s, records an 'Old Engine' in Park Estate to the south next to the boundary with Felling estate; and another map dated before 1770 records an engine at Claxton's Colliery. It is odd that Saltmeadows appears in Brown's list but no map exists of an engine in this part of Gateshead Park.[196]

In about 1766, Brown estimated that there was 42,000 tons of coal left in Park which could be won with an engine costing £1,800. Aubonne Surtees and Co. leased the colliery in 1775 and the viewer John Donnison wrote to Henry Ellison the lessor in 1787 that 'there are two complete Fire Engines erected on Park Colliery, with cylinders of 61 inches diameter each, which have won the upper main coal seam at a depth of 28 fathoms below the Tyne level'.[197] Another report, dated August 1791, claimed that the

---

[195] NEIMME: GA/2/37; Forster 1/4/157; LBWB I p. 262.
[196] NEIMME: Watson 29/1; NRO-SANT-BEQ-09-01-03-15; NRO-SANT-BEQ-09-01-03-01.
[197] Manders, F.: 'History of Gateshead' p.57.

consequences of stopping the engines draining the High Main at Park would be 'to preclude the possibility of Working the collieries of Felling and Heworth to profit' and would have consequences for Byker St. Anthony's, Walker, Bigges Main, Heaton and Wallsend. The strategic importance of Gateshead Park was reinforced in 1823 when a new pumping engine was built at Friars Goose with a 180 HP engine capable of drawing 1,444,800 gallons a day from the High Main seam. This engine was funded by a consortium of collieries in the Tyne Basin. In 1841, a 70 HP engine was added to drain the Low Main seam. The breakup of the consortium in 1851 was followed by a flooding of the Tyne Basin and the closure of the major collieries: Heaton in 1852, Wallsend in 1854, Willington in 1856 and Bigges Main in 1857. The ruins of the pumping engine remain on the banks of the Tyne near Gateshead International Stadium.

Newbottle In 1733, John Nesham took over the Earl of Scarborough's colliery (also known as Houghton Colliery) including the existing fire engine. A view by Amos Barnes, William Newton, Edward Smith and his son dated December 1733 records the engine at work and suggests a drift to drain East Rainton Moor Colliery. In 1740 the cost of keeping the engine was £400 p.a. By the mid 1740's the High Main coal was becoming exhausted and a new winning was attempted further north. Amos Barnes reports chart the progress of the work between October 1746 and May 1748 which involved moving the existing engine to a new site. The charge for 'finding New Materials that are wanting and drawing and repairing the Old Engine Materials and fixing them in the Newhouse' was £600. It was also recommended that 'a Smith's Shop should be immediately built a little to the South East of the New Engine in order to build a Boyler and repair the other Materials wanting for the said Engine'. By May 1748, it was reported that 'The Engine may be erected in 12 weeks time provided plates be immediately got and a Boyler Set forward with all the Old Materials Ledd to the New Engine and repaired'. The new colliery opened in 1750. Brown list records two engines at Newbottle and his map (NRO, SANT- BEQ 09-01-03-52) shows two old engine pits at Newbottle which was probably the site of Nesham's engine. The map also shows an engine on the Houghton Colliery and one at Morton further west. [198]

Elsdon Elsdon was a landsale colliery in a very remote part of north Northumberland between Otterburn and Rothbury. Peck provides a quote for a bob gin in 1723 for John Liddell and Co., not a fire engine. He provided a similar quote for Hartford Colliery. [199]

Eldon This was a more important landsale colliery near Bishop Auckland in south Durham where Peck was also involved but not in building fire engines. [200]

Jarrow The estimate by Amos Barnes, Richard Peck and William Dryden for the new winning at Blackett's High Heworth Colliery in 1742 records that the engine was used during the sinking of the shaft: [201]

Building Engine House and Engine	£1,200
Sinking Engine Pit	£522
Keeping Engine 90 weeks while pit is sinking at £4 p. fath	£360
Coals burnt 90 weeks	£432

This engine is shown on John Watson's map of Jarrow Colliery dated 1750 (fig.59) located at High Heworth.
Brown records two engines at Heworth, one with 52" and the other with 72" cylinders. The smaller engine, the earlier of the two, is marked on the map.

---

[198] NEIMME: Forster 1/4/228; Forster 1/4/59; Forster 1/4/116, 118,121, 122, 138; Forster 1/4/116, 118,121, 122, 138.
[199] NEIMME: Peck p.26 and 27; Peck p. 37.
[200] NEIMME: Peck p.73.
[201] NEIMME: Forster 1/4/82.

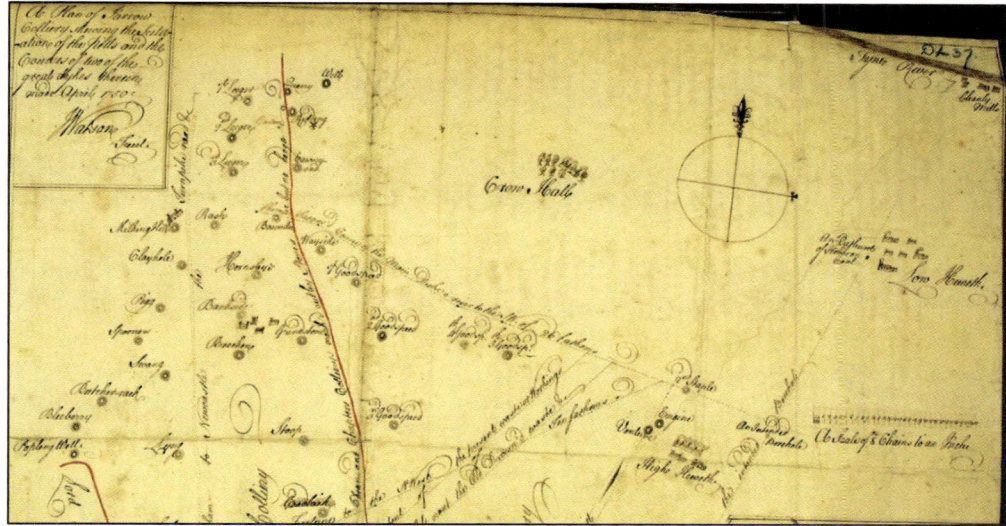

NEIMME: Watson 30/7

Fig. 59: Blackett's Jarrow Colliery in 1750

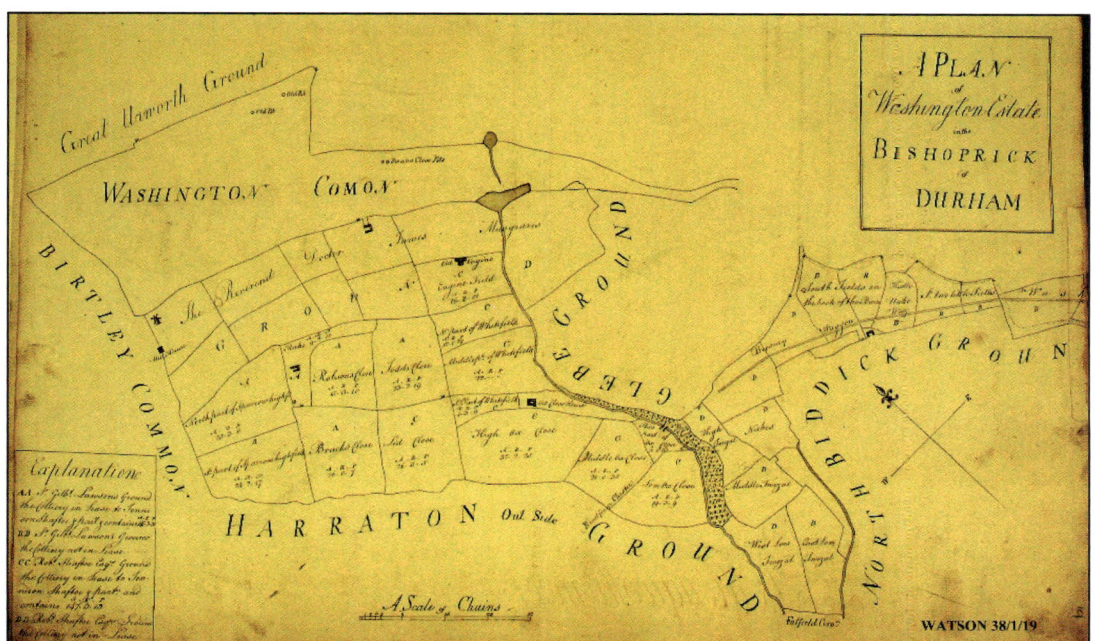

NEIMME: Watson 38/1/19

Fig. 60: Washington Estate circa 1750

Washington	The site of Beighton's engine of 1718 is recorded at the north end of the Engine Field as 'Old Engine' in the Oxclose district of Washington (fig. 60). A later map, dated circa 1764, of the same area shows Shafto's colliery and the Engine Field, which by then housed a later engine, which was probably the engine on Brown's list. In 1931 a local resident reported that the walls were 10 yards long and 18 inches wide. Huddleston informs us that, in 1775, Brown built a double engine house at Washington for Russell and Wade's colliery situated further north but only one engine was needed.[202]

---

[202] NEIMME: Buddle 7/11; NEIMME Trans 82, p 529.

**Appendix II: Family Connections – The Browns, Newtons and Watsons
by Andrew Curtis and Les Turnbull.[203]**

Hidden within the parish records of St. Andrew's Church at Heddon on the Wall are the baptism, marriage and burial records of the Brown family. In examining these documents, the student experiences all the thrill and frustration of family history research. William Brown II (1717 – 1782) of Throckley Fell, the subject of this book, was baptised at Heddon parish church in 1717 and buried at the same church in 1782. He was married to Mary Smith at Morpeth in 1741 and they had seven children two of whom died in childhood and are remembered by a plaque on the south east outer wall of the church. Their eldest son, William Brown III (1743 – 1812) married Margaret Dixon at Stamfordham in 1770 and they followed the custom of naming their eldest son after his father. The couple had eleven children three of whom were called William Brown: their first William (1772 – 1778) died in childhood, their second William (1781) died in infancy and the third (1785 – 1813) only survived to early manhood. William Brown II's father, William Brown I, was buried in Heddon churchyard in 1746.

There is an entry in the Heddon church register which reads 'William Brown Senior of Heddon on the Wall Pit House was buried in the churchyard April 29th 1715' and it is likely that this man was William Brown II's grandfather. However, there is another reference to the Browns of Heddon Pit House in 1730 which strikes a discordant note: 'William, son of John and Isabel Brown of Heddon Pit House, baptised privately'. An alternative explanation is that there were two families of William Browns, one living at Heddon, and our family living at Throckley. In a petition to the vicar and church wardens of Heddon dated 22nd March 1777 regarding a burial place for his family on the south side of the church, William Brown II refers to 'a piece or parcel of ground in the church yard there where the bodies of several persons of his own family have, alone, as it is believed for these sixty years past and upwards been interred'. The sixty years past would fit with the 1715 burial in the register. However, Brown is a common surname in this part of Tyneside and it cannot be assumed that others of that name are directly related. When she inherited East Denton, Mrs Montagu refused to employ John Roger's untrustworthy servant called William Brown but following the death of William Newton in 1763, she was happy to replace her viewer with William Brown of Throckley. Was John Rogers' servant the William Brown of Heddon Pit House or was he the William Brown who captained the keelboat which supplied wood to the Throckley waggonway? Clearly there was no shortage of William Browns in the area.

William Brown's mother was Anne or Agnes Watson, born in 1693, the daughter of Lewis and Jane Watson of Close Lea in Heddon. She had a brother John baptised at St. Andrew's Heddon in 1690. Was this the same John Watson whose marriage to Sarah Newton on 10th April 1715 is recorded in the parish records of All Saints Church in Newcastle; and was this man John Watson the mariner who baptised his son, John Watson II at the same church on 12th October 1729? Unfortunately, Watson is also a common surname in the area and we can't be sure. However, we do know that John Watson II (1729 – 1797) was apprenticed to his uncle William Newton of Burnopfield, one of the most distinguished viewers of the mid eighteenth century, who managed the great collieries of the Windsors and Lambtons amongst others. Watson view book records his work in the service of his master at most of the major collieries in the

---

[203] We are grateful to two people who have kindly given us access to their research into their family histories: Anne Willoughby who has worked on the records of the Browns and Jonathan Peacock who has researched into the Watson family.

Tanfield area, on Tyneside, at Plessey and at Bedlington.[204] Upon William Newton's death, Alice Windsor, like Elizabeth Montagu and Matthew Ridley, invited Brown to be Newton's successor and John Watson II became part of Brown's consultancy. He was associated with Willington Colliery managed by George Johnson another member of the consultancy; and his grandson John Watson III (1786 – 1847) became Johnson's apprentice. The Watsons lived at Willington House after the death of William Brown and became shareholders with George Johnson in Heaton Main Colliery.

As a result of his successful career as a viewer, in later life William Newton owned a colliery at Beamish and lived in the splendour of Burnopfield Hall in sight of the Western Way, one of the great railways of the coalfield which carried Windsor's Pontop and Ridley's Tanfield Moor coals down to Derwenthaugh. There is a record of a William Newton being baptised at Heddon in 1695 and his parents are recorded as John and Elleanor. A John Newton and Elleanor Ferrer were married at Holy Cross Church Ryton in 1672. However, William Newton is also a fairly common name. Eneas MacKenzie's records in his History of Durham (1834) that 'On a flat stone at Tanfield – William Newton departed this life Nov. 21st 1763 aged 63 years'. This definitely referred to the famous engineer and since it states that he was born about 1700 he is not the Heddon man. William Newton's family had a troublesome history: his daughter Hannah had the misfortune to be married to the Irish fortune hunter Andrew Robinson Stoney, later to become infamous as Stoney Bowes, the husband of Mary Eleanor Bowes, the Dowager Countess of Strathmore; and his son was a reprobate who squandered the family fortune and became a bankrupt in 1773.

The genealogy of these three families is not easy to trace. William Newton and John Watson were certainly related but whether William Brown was related to these families is only speculation. Dunn records in the 'History of the Viewers' Sam Haggerston's recollection that William Newton recommended William Brown for the post of viewer at Throckley seasale colliery and this may be significant. What is certain is that the world of mining engineering in the eighteenth century was a close knit community and business relations were often cemented through family ties. Skills were acquired, training given, connections made and business opportunities shared through the family network. For a successful career as a viewer discretion was necessary and this was probably best achieved within the close confines of the family firm.

---

[204] NEIMME: Watson 2/4 is a collection of extracts from John Watson II's view books. Someone has written a frontispiece which states that John Watson was William Newton's cousin which is incorrect. Unfortunately, this has been repeated in subsequent histories. While serving as Newton's apprentice, John Watson II produced three beautiful maps now in the Mining Institute, one of the Earl of Carlisle's estate at Longbenton in 1749 (20A/12), another of Pontop Estate in 1750 (31/19) and the third of Bushblades (31/14) also in 1750. These are but a small part of the collection of papers donated to the Institute by William Watson in 1863 which have played a significant part in the writing of this book.

Appendix III: The Archaeology of the Eighteenth Century Coal Industry

Scattered throughout the previous chapters of this book is a great deal of information about the archaeology of the Northumberland and Durham coalfield in the eighteenth century. However, it seemed appropriate to pull the information together and provide a guide to the tools and machinery which were used at this time in the extraction of coal at the coalface and its transport through the mine, to the surface and along the waggonway to the staiths for shipment.

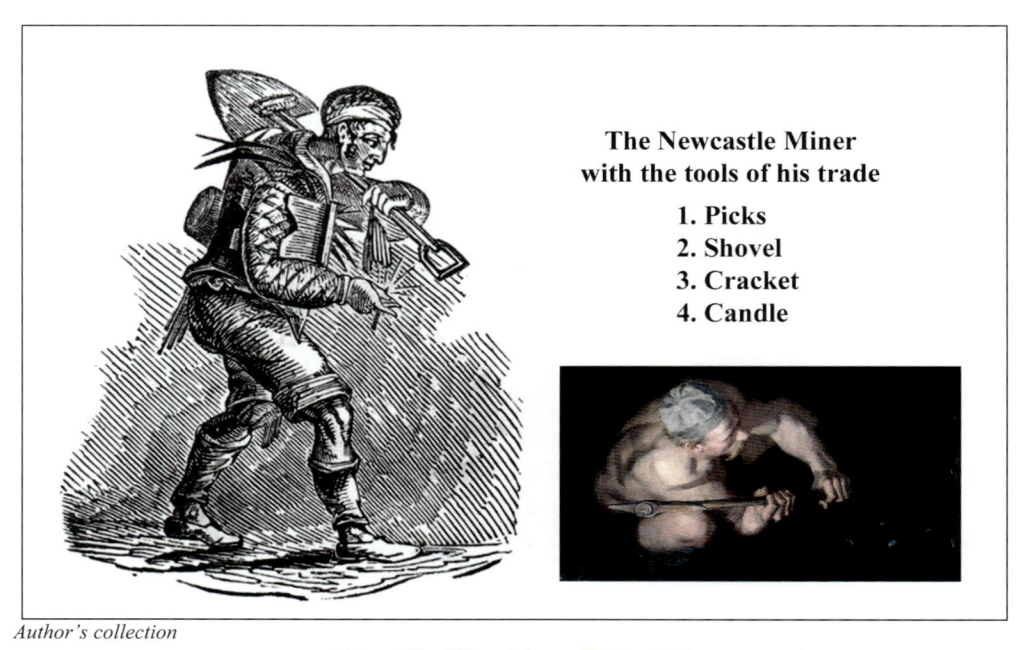

Author's collection

Fig. 61: The Newcastle Miner

In the eighteenth century, the Newcastle miner was a man of independent spirit who worked by candlelight underground using his own picks sharpened by the colliery blacksmith. Using his own shovel, he loaded coal into corves usually made of hazel wands which held about six hundredweight. These were made by the corver who was often an independent tradesman. At Bigges Main he lived in the large house at the end of Blue Row. Sir John Clerk visited Alderman Richard Ridley's colliery at Byker in April 1724 and made a drawing of a corf in his diary (fig.62). He described how 'the colliers that work belowe are payed so much for each corfe and each of them sends up their owne marked by a Talley', a practice that continued until the end of mining.

National Archives of Scotland GD18/2106

Fig. 62: Extract from the Journal of Sir John Clerk April 1724

The corves were placed on a sledge and moved from the coalface to the bottom of the shaft, known as the eye of the colliery, by putters. In the shallow collieries with multiple pits this was usually a distance of about 100 yards. As collieries became deeper and the distances between the coalface and the shaft became greater, other forms of transport were introduced. By the second half of the century wooden trams were in use such as the one illustrated by the Frenchman Morand in 1768 (fig.63); and by the end of the century, the major collieries had an extensive network of railways underground which used horses and included inclined planes. Hair's watercolour of Walbottle Colliery in the 1830's shows scenes which would have been familiar to miners working in Brown's collieries at Willington and Bigges Main in the late eighteenth century (fig. 64).

NEIMME: Morand
Fig.63: A wheeled sledge or tram

NEIMME: Hair
Fig. 64: Underground at Walbottle Colliery

The mining engineer had to address the problems of draining, lighting and ventilating the colliery. We have already encountered many pumping engines in the previous chapters from the small engine at Tanfield in 1715 to the massive double engines at Walker and Shiremoor in the 1760s. Shortly after William Brown's death, his son opened up East Benton Colliery in 1786 together with George Johnson and Matthew Bell. The principal sinker, Matt. Tubman, was presented with a medal for his services which has an illustration of what was in all probability one of Brown's engines. The partnership built the village of Bigges Main to accommodate the workforce. The first edition of the Ordnance Survey Map shows the location of the engine shaft at the end of High Row. Like the company's other mining village at Willington Square, Bigges Main comprised mainly terraces of single storey cottages with the waggonway between. Note the profusion of allotment gardens. The houses were very attractive accommodation in their day and both villages lasted well into the twentieth century. All of this enterprise is now lost beneath the greens of Wallsend's Centurion Golf Course.

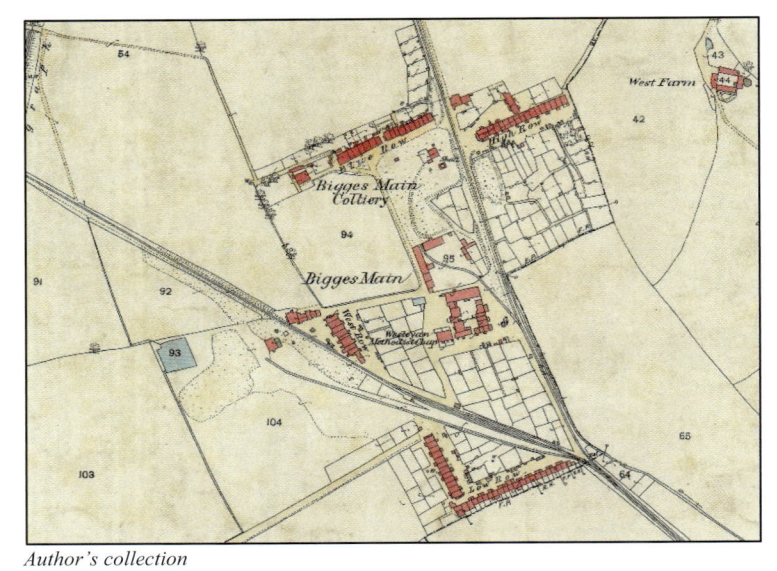

Author's collection
Fig. 65: Bigges Main Village in 1858

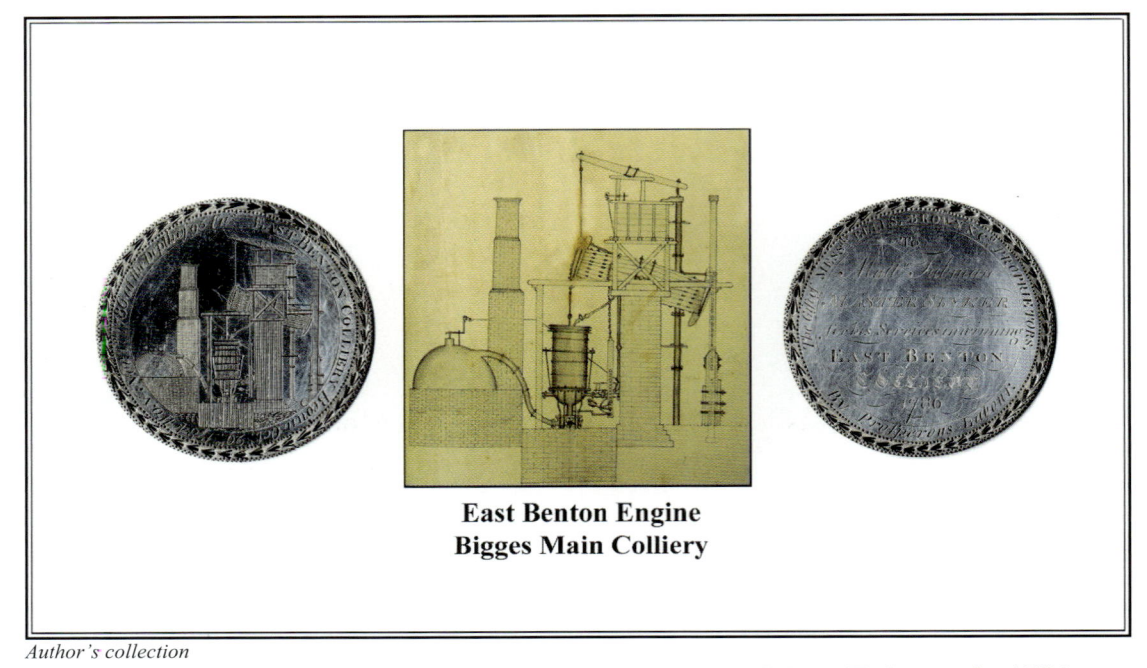

**East Benton Engine
Bigges Main Colliery**

Author's collection
Fig. 66: East Benton Sinker's Medal presented to Matt. Tubman in 1786

Tallow candles were the principal source of lighting underground in the eighteenth century. These were usually carried in a small tin known as a midgy lamp or fixed to a pit prop near the workplace. Generally, the hewer was required to buy his own lighting to use at the coalface but the company paid for lighting elsewhere. In the larger collieries oil lamps were in use, the oil being a bye-product of the important whaling industry on the River Tyne. In about 1730, Carlyle Spedding invented the steel mill as a safer way of working the gassy pits around Whitehaven. According to Dunn, the steel mill was first used in the North East at Fatfield Colliery in 1763. This early safety lamp was strapped to a boy who turned a geer wheel which operated a steel disc at speed. By placing a flint against the steel disc a shower of sparks was created which provided enough light for about half a dozen men to work by. The sparks changed colour in the presence of methane gas which provided an early warning. However, the apparatus was not completely safe as John Buddle demonstrated after several accidents at Wallsend Colliery. It was not until the development of the safety lamp by Stephenson and Davy in 1815 that the dangers of a gassy pit were lessened.

Fig. 67: The Steel Mill

Before mines became extensive, changes in barometric pressure were sufficient to provide a circulation of air within the pit: a rise in pressure would drive the air into the working and a fall draw it out. As mines became larger the problem of ventilation became more acute. In the seventeeth century, a fire basket was suspended down the shaft to create convection currents to circulate air; and in the larger mines an additional ventilation shaft was sunk for the same purpose. At Byker Colliery in 1724, Sir John Clerk saw the 'methode which was used to drain ill aire out of Alderman Ridley's works and likeways to carry a mine a great way without the expense of letting down a shaft'; and he provided an illustration (fig.67) in his journal. 'At the mouth of the shaft **A** he made a furnace **B** & at the aire hole **C** he fixed a pipe **D** of timber which was let down the shaft **A** and by degrees convoyed to the end of the mine or wall face at **E**'.

Fig. 68: The Ventilation Apparatus at Byker Colliery 1724

The use of wooden boxes had a long history and as late as 1848 Matthias Dunn could write that 'until lately it was quite common to see collieries dependent for their ventilation upon a wooden box 12 or 14 inches square, sometimes led to the fire engine as a substitute for a proper furnace, and sometimes operated by a small revolving fanner, worked by a steam engine'. The fire engine was used to work two ventilation fans at Walker Colliery in 1769. The problem of replacing foul air and noxious gasses increased as collieries became larger in the later eighteenth century and the method of coursing the air around the workings by using convection currents became common practice. At Wallsend Colliery, John Buddle devised the technique of dividing the colliery into districts to improve this technique. In Buddle's diagram below, the shaft **a** is divided by wooden bractices into two downcast shafts, **b** and **c**, and one upcast shaft **d** near the furnaces **ff**. The air is directed around all the workings by means of bracticing and trap doors. Incredibly, the responsibility for operating the doors, so essential for operating the ventilation system, was in the hands of young children – the trapper boys!

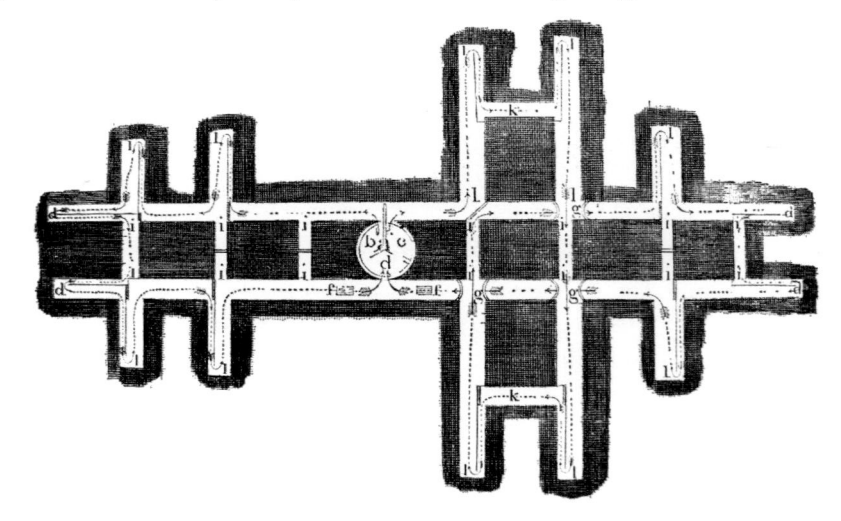

Fig. 69: Buddle's diagram showing how the air was directed or coursed around the workings of the colliery [205]

Clustered around the shaft at the bank head was the ventilation, pumping and winding gear. In the first half of the century, the cog and rung gin was the common form of winding machinery (fig.70). Power was provided by horses walking in a large arc around the shaft. They generally worked an six hour shift.

NEIMME: Taylor

Fig. 70: Cog and Rung Gin

---

[205] In his report into the disaster at Felling Colliery in 1813, the Reverend John Hodgson provides a detailed description of how the system of coursing the air operated.

The cog and run gin was replaced by the whim gin (fig.71). By this method the winding machinery was well clear of the shaft.

NEIMME: Taylor

Fig.71: The Whim Gin

The problem with both these machines was the expense of using teams of horses and colliery engineers turned to other methods of winding – at first water power and then steam power. Figure 12 shows Menzie's engine of the 1750s and by the end of the century waterwheels were in use at many of the major collieries on Tyneside including Bigges Main, Wallsend, Walker and Felling. The early experiments with steam power for winding at West Longbenton and Hartley have been noted above but, although the first of Watt's engines in the region was used for pumping at Byker in 1778, the first record of a Watt engine for winding is at Benwell in 1795, which was followed by others at Walker, Wallsend, Cowpen, Heaton and Willington in 1796. The bank head at Wallsend 'C' Pit, which was opened in 1790 shows a steam winding engine in operation. Note the flames to burn off the methane gas, which was particularly troublesome at this mine; the corves being dried around a fire; and the screens which are believed to have first been introduced by William Brown at Willington.

NEIMME: Hair

Fig. 72: Wallsend 'C' Pit

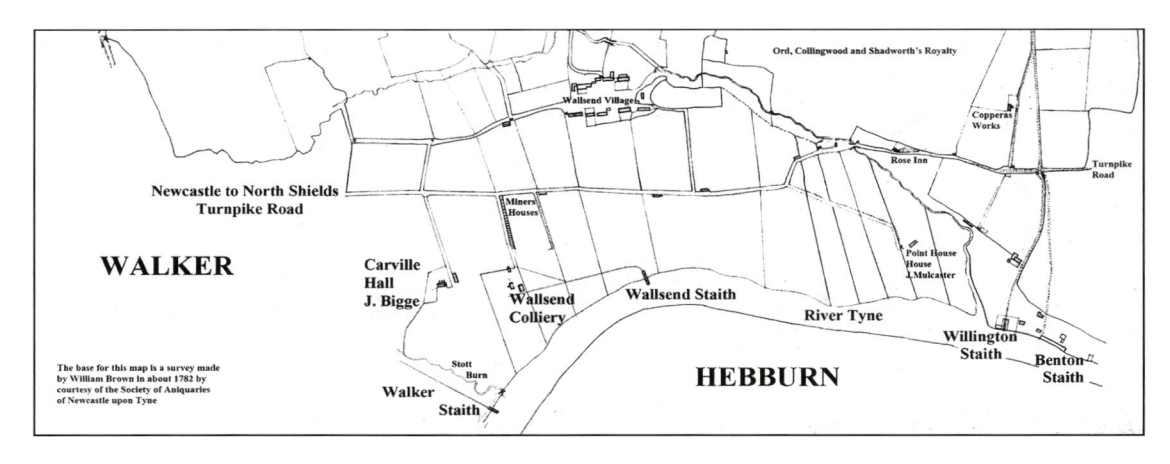

Author's collection

Fig. 73: Wallsend circa 1782 showing the beginnings of Wallsend Colliery

The only place where it is still possible to see the remains of an eighteenth century coalmine is at Wallsend where the excavations on the site of Hadrian's Wall near the fort of Segedunum undertaken in 1997 revealed the 'B' pit (fig.74). The location is shown on the map from Brown's collection dated 1782 (fig.73). A branch line leads from the 'B' Pit to join the line from the 'A' Pit, now the course of Buddle Street, en route to the staith. This railway was the work of William Brown and his assistant William Gibson.

Author's collection

Fig. 74: Remains of Wallsend 'B' Pit – Possibly Brown's Work

The concrete slab in the top left corner is the cap for the 'B' Pit shaft. In the foreground are the remains of the engine house and the circular brick building is one of the boiler houses. It is a sign of our times that a considerable sum of money has been spent on reconstructing Hadrian's Wall and the Roman bath house when the industrial remains are left with little interpretation. Yet these humble monuments were very important for the history of Wallsend and in terms of the region's place in the history of the world, the mining industry was unquestionably more important than the Roman frontier.

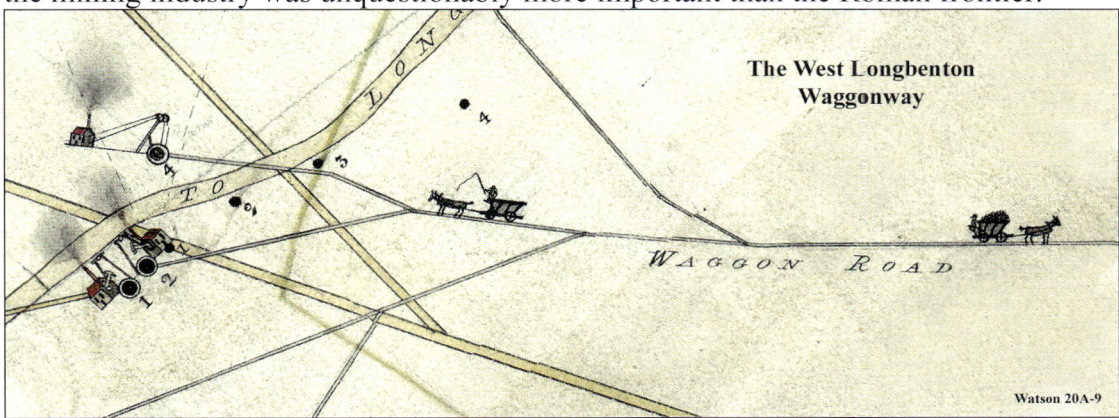

NEIMME: Watson 20A-9

Fig. 75: The Waggonway

This section from John Watson's beautiful map of West Longbenton Colliery drawn in the summer of 1749 is a reminder details of waggonway design which have been discussed fully in Chapter Six. The waggonways were the means of efficiently moving the coal from the collieries to the waterside for shipment. Care was needed in handling the coal because it was friable and easily broken when the market required large round coals. In the early eighteenth century the shoot was in general use. Taylor's drawing shows the chaldron waggons delivering coal to the staith, a large wooden building where coal could be stored to prevent it weathering. The coals were dropped down the large shoot into a keelboat. One is showing leaving with its crew of three on the journey downriver to the colliers moored at Shields (fig.76). In the later eighteenth century, the principal coal shipment points were nearer the mouth of the river where the colliers could moor alongside the staith although they generally needed to be topped up in the deeper waters by the keelmen (fig.78).

NEIMME: Taylor

Fig. 76: The Coal Shoot

the ground. The eighteenth century mining landscape between Walbottle and Callerton is uncovered by this twenty first century technology. Of particular interest is the network of waggonways serving the numerous pits on the extension of William Brown's line from Walbottle northwards to Callerton. The great cutting, which has been partly destroyed by the A69, is prominent. At the north end, Cutend Junction, one branch of the railway leads to the Tommy Pit and beyond while the other leads past the Betty, Andrew and Brass pits to the Dolly Pit where George Stephenson worked as a youth. On this eastern fork is another junction leading northwards in the direction of the Jockey pits where there are traces of another waggonway, or perhaps a wain way.

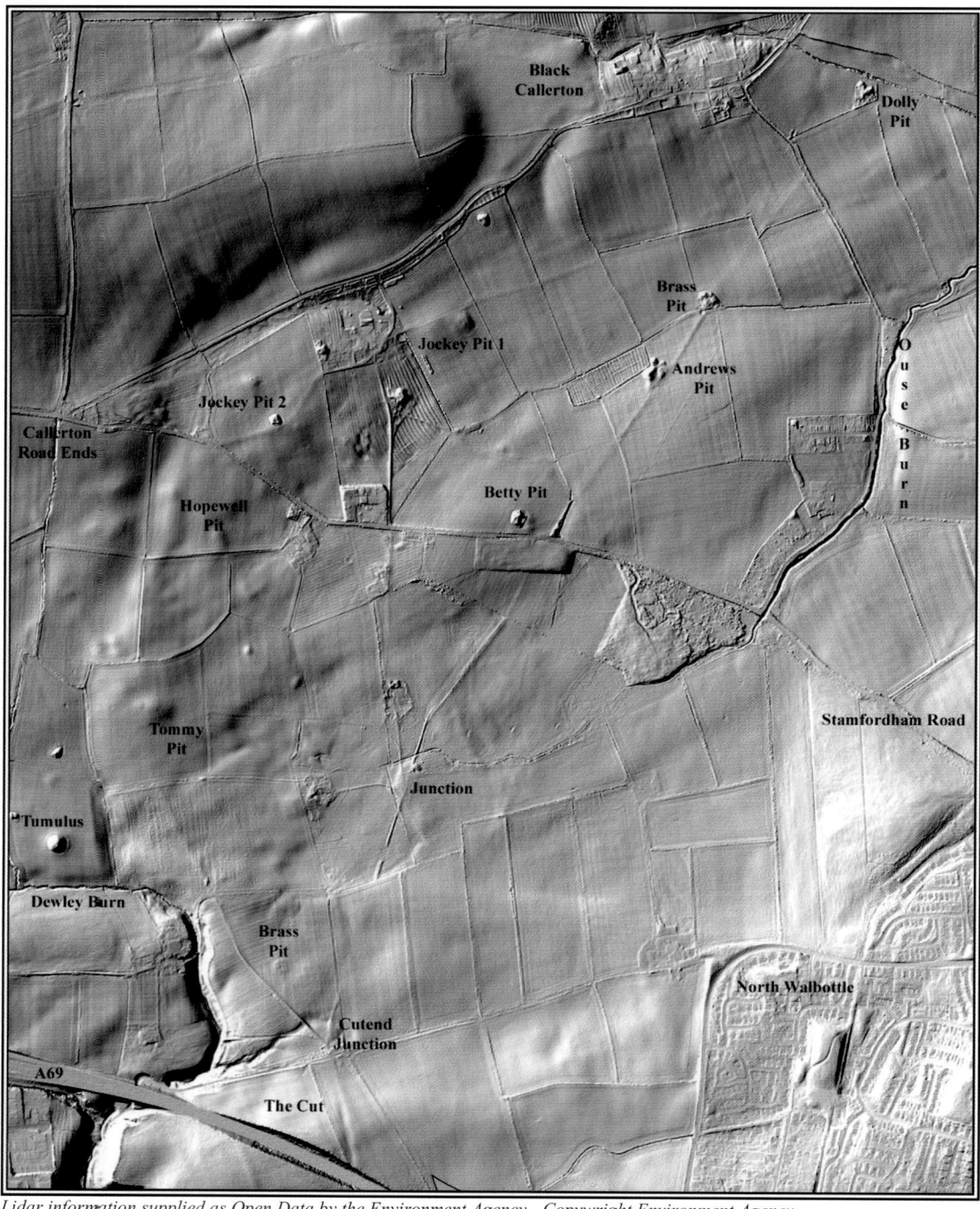

Lidar information supplied as Open Data by the Environment Agency - Copyright Environment Agency.

Fig. 77: Lidar Survey revealing an Eighteenth Century Mining Landscape

Bibliography

This book is largely based on primary sources and the manuscript collections used in this study are referred to in the copious notes which accompany the text.

Flynn, Michael W. 'The History of the British Coal Industry Volume 2: 1700 – 1830 The Industrial Revolution (Oxford 1984) is the standard history.

Clothier, Alan C. 'Before the Blaydon Races' (Melrose Books 2014), contains much interesting information particularly on the later history of the collieries leading their coals to Lemington. However, he exaggerates the importance of Wylam Colliery in the eighteenth century.

Williamson, Bill 'Class, Culture and Community – A Biographical Study of Social Change in Mining' (London 1982) enriches our understanding of the area although it refers to a later period.

Index

General

Absconding workers	58-59	drainage by waterwheel	81
'a fighting trade'	31-32		
annual bond of miners	42	East Denton treat	56-57
apprenticeship	41		
		false rail	67, 70
batteries	65	fitter	34
board	52	fother	45
bob gin	81, 83-85		
Boulton and Watt engine at Byker	92	Gateshead Park	106-108
Boulton and Watt engines at Walker	95	gradients on waggonways	74
Brown, William			
Benwell Colliery engine	19	Haggerston, Sam (recollections)	97, 99-100, 102
birth (1717)	21	Heaton engine	84-85
Brown, Lancelot 'Capability'	23	horses (and their costs)	79-80
builder of steam engines	24		
coat of arms	28-29	Lemington community	77
concern for suffering	25	Lemington staithes	15-17, 77
connections with iron works	85	loss of workers to other pits	50-51
connections with Newton family	22, 110-111		
connections with Watson family	22, 110-111	Newcomen engine	26-28, 82 et seq
death (February 1782)	98		
dispute with Mr. Humble (1752)	48	polluting the Tyne	81
early debts	21, 50	press gangs	34, 36
eldest son	29		
exploitation of the Tyne Basin	87-88	Ravensworth engine (1570)	81
family	21-23		
Hartley Colliery (1759)	90	Seaton Sluice	90
interest in antiquities	23	Seven Years War	34
inventor	23-24, 26	staithman	78
largest pumping engines	88	standard railway gauge	9
list of Newcomen engines	99-109	strata in the Tyne Basin	87
mother, Anne Watson	110		
personal life	23	tenants' obligation to lead coals	60
Shiremoor Colliery (1767)	90-91	tender for a waggonway (1788)	64-66
silver punch ladle	94	Throckley Bridge	62-63, 69
steam engines	81 et seq	Throckley Waggonway	47, 62-64, 66-78
Survey of Brunton (1752)	13	trade in coal	10-11
Wallbottle Waggonway (1771)	66-69	trade in coal	10-11
Throckley Waggonway	15-16, 66-78	trade in coal	10-11
tobacco box	29		
travel to Scotland to seek men	51	underground trespassing	93
West Brunton Way	13	use of latest technology	20
Willington Waggonway	73, 80		
Byker colliery engines	91-92, 104-105	vends in River Tyne (1753)	32
		viewer	37 seq
Carville excavations in 2013	73, 80		
chaldron waggon	74-75	waggoner's bond	77-78
Coalbrookdale (Abraham Darby's foundry)	85	waggonways	12, 64-80

coal factors	34-35	waggonway traffic besides coal	78, 97
coal trade in mid eighteenth century	30-31	waggonway worked only in daylight	78
contract for William Cramlington	59-60	Walbottle Colliery village	72
crease	66, 71	wash hole	72-73, 80
cutts	65	workforce needed at a colliery	54, 56

People

Allen, John	87, 93	Allison, Ralph	39
Archdeacon, William	17, 39	Allcock	29, 45
Baker, John	18, 19	Barkas, William	29, 34, 45, 48, 49
Barnes, Amos	21, 37, 41, 49, 83, 84, 88	Barnes, Jonathan	37
Beaumont, Robert	13, 16, 18, 19, 50, 51	Bedlington, Chris.	9, 13, 19, 20, 33, 37, 41, 63, 81, 87, 90
Bedlington, William	9, 37	Beighton, Henry	83, 101
Bell, Matthew	28, 29, 63, 71, 90, 95	Bewick, Colonel	17,
Biggins, John	35, 36	Blackett, John	17, 18, 50, 51, 63, 64, 76
Blackett, Sir William	10, 12, 78, 88	Boag, John	40
Blenkinsop, John	79, 80	Brandling, William	66, 93, 98
Bridgewater, Duke of	17	Brindley, James	17, 89, 98
Broade, Thomas	89	Brown, Agnes	22
Brown, John	22	Brown Lancelot	22, 23
Brown, Mary	22	Brown, Richard	22, 37, 41, 42, 98
Buddle, John	38, 58, 66, 71	Burrell, John	24, 50
Carlisle, Earl of	9, 12, 22, 33, 85	Carter, Elizabeth	39, 80
Chapman, William	38, 42, 93	Chicken, Edward	57
Claughton, George	49, 88	Clavering, Lady Jane	9, 29,
Clavering, James	21,	Clerk, Sir John	5, 80, 82, 91
Cotesworth, William	21, 83	Cramlington, William	13, 59, 60, 66
Darby, Abraham	10, 49, 85, 89, 98	Davison, Morton	23,
Delaval, Sir John Hussey	89, 90	Donnison, Thomas	71,
Dunn, Mathias	23, 99, 100, 101	Edington, Robert	28, 29, 37
Elliot, George	58	Emerson, William	38,
Errington, Anthony	66	Fiennes, Celia	10-12,
Forster, Jonathan	7	Forster, Henry	77
Forster, Richard	64, 65, 66	Forster, Robert	67
Gibson, John	4, 41, 61, 70, 80, 97	Gibson, William	41, 42, 43, 44, 59, 87, 95, 96
Goldney, Thomas	88	Green, George	45
Hackworth, Timothy	7, 9, 20,	Haggeston, Sam	23, 93, 96, 98, 99, 100
Hartley, Leonard	24-28, 85, 87	Hamilton, Duke of	22, 24, 50
Hawthorn, Robert	8, 9,	Hedley, William	7, 8, 17, 20
Hesilrige, Sir Robert	12	Hodgson, Rev. John	9,
Hull, Jonathan	86	Humble, John	15, 31, 35, 48, 77, 87, 90
Hutton, Charles	38, 58	Jars, Gabriel	94, 95
Johnson, George	28, 41, 45, 87, 93, 95	Khedave of Egypt	8, 58
Lambe, Joseph	70	Lambton, Henry	83
Lawson, Sir John	92	Liddle, Sir Henry	37, 81, 83, 89
Locke, William	8	Locke, Joseph	8
Lowther, Lord	10	Masterman, Henry	16, 29, 33, 34, 35, 45-51, 66
Menzies, Michael	23, 38	Milbank, Sir Ralph	94, 95
Montagu, Charles	19, 20	Montagu, Edward	16, 17, 18, 21, 37, 38, 39, 40, 54, 90
Montagu, Elizabeth	9, 21, 39, 40, 55-6, 62, 80, 90, 96	Nesham, John	84
Newcomen, Thomas	82,	Newton, William	10, 39, 40, 49, 88, 92
Northumberland, Duke of	10-15, 22, 50, 59-66, 71-81, 90	Oliver, William	87
Ord, John	9,	Oxford, Earl of	5, 81
Oxley, Joseph	23, 86	Peck, Richard	9, 13, 19, 20, 21, 22, 37, 83, 84, 88, 93
Peck, William	14	Peareth, William	41, 93
Potter, John	83, 84	Portland, Duke of	23
Rawlings, Jonathan	25	Ridley, Matthew	37, 54, 92
Ridley, Richard	82, 83, 91	Rogers, John	9, 15, 16, 40
Rutter, Christopher	21, 22	Savery, Thomas	82,
Scarborough, Lord	83, 84	Smeaton, John	85, 86, 89, 98
Smith, Edward	40	Smith, Jonathan	39
Spedding, Carlisle	10, 22-31, 38, 47-9, 74-5, 88, 101	Spedding, James	52, 89
Spedding, John	88	Stephenson, George	7, 9, 20, 58
Stephenson, Robert	8	Stukeley, Dr. William	5, 80, 81
Taylor, Thomas	78	Thomas, William	38, 39, 56
Thompson, Isaac	15, 16, 49, 50, 53, 88	Thompson, Jonathan	50,
Walton, Nicholas	10, 37, 40, 100	Watt, James	86, 92, 95
Watson, John	12, 95	White, Matthew	19,
Wilkinson, John	12, 88	Wise, John	86
Woodhouse, Francis	23	Wright, John	18, 39

Collieries

Lists (Vends)	32, 100, 102, 104		
Aston	42	Banklands	59
Benwell	9, 18, 19, 24, 81	Biddick, North	83
Biddick, South	83	Bigges Main	65, 69, 70, 79, 95
Birtley	40, 41, 71	Black Close	23, 75
Blaydon	29	Bo'ness	22, 24, 50
Brunton	12	Bushblades	9, 84, 105-6
Byker	9, 37, 82-92, 101, 104	Charterhaugh	38, 83
Coale Fell	9	Coxlodge	19
Denton, East	16, 18, 20, 30, 40, 55-6, 80-1, 97	Denton, West	18, 88
Elswick	19, 20, 82	Farnacres	83
Felling	9, 81, 93	Flatts	5, 83
Ford	93	Friar's Goose	92, 100
Frystonhall	42	Gateshead Park	83, 101, 106
Hartley	9, 85, 86, 89	Heaton Banks	9, 37, 83, 84, 85, 87, 88
Heaton Main	71	Heddon	16, 29, 45, 45, 52, 62, 64
Heworth	42	Holiwell (Newbiggin)	13-15, 17, 33, 47, 50, 51, 52, 70
Hucknall	23	Jarrow	84, 88, 108-9
Jesmond	9, 20, 88	Kenton	9, 12, 13, 18
Killingworth	58	Lemington	16, 18,
Longbenton West	9, 37, 38, 58, 80, 85-89	Longbenton East	96
Low Fell	93	Lumley	81
Newbottle	83, 108	Newburn Moor	14
Newark (Fife)	72	Norwood	82
Oxclose	82, 83	Ravensworth	81, 89
Shiremoor	28, 31, 90	St. Anthony's	92, 93
Stella	29	Tanfield Lea	82, 84, 105
Throckley	16, 29, 30, 33, 40, 44, 45, 46, 51, 52, 61, 62, 64, 70, 96		
Walbottle	17, 20, 33, 42, 54, 59, 63, 72, 76, 77, 81		
Walker	25, 90, 93, 94, 100	Walker Hill	9, 87, 92, 93
Wallsend	43, 87, 95	Warnell Fell	23
Washington	22, 23, 44	Whorlton	15, 22, 50
Willington	28, 46, 70, 79, 95, 96, 97	Winlaton	9, 29,
Workington	27	Wylam	17, 33, 50, 51, 76

NEIMME: Taylor

Fig. 78: A Collier loading at a Spout

This publication has been made possible because of the support received from:

The Alan Auld Group Ltd

The Newcastle Centre of the Stephenson Locomotive Society

The North of England Institute of Mining and Mechanical Engineers

and the following individuals:

Alexander Appleton
Russell Benson
David C. Bell
Don Borthwick
Simon Brooks
Chris Calver
John T Crompton
Andrew Curtis
Mick Eyre
Eric Fisher
Charles Fleming
Paul Gailiunas
Steve Grudgings
Peter Hemy
Gordon Hull
John Irving
David Kitching
David McAnelly
Colin Mountford
Derek Newton
L. C. Paul
Malcolm R. Paul
Dr Ian David Pearce
Thomas J. Preston
David Sharp
David Shepherd
Phil Thirkell
David Tyson
Dr Eric Wade
Alan Williams
Janet Whiting
Tom Yellowley

A very sincere thank you for making this publication possible – Les Turnbull